Composite Materials: Science and Engineering

Composite Materials: Science and Engineering

Douglas Holliday

WILLFORD **P**RESS

www.willfordpress.com

Published by Willford Press,
118-35 Queens Blvd., Suite 400,
Forest Hills, NY 11375, USA

ISBN: 978-1-64728-340-7

Cataloging-in-Publication Data

Composite materials : science and engineering / Douglas Holliday.
    p. cm.
Includes bibliographical references and index.
ISBN 978-1-64728-340-7
1. Composite materials. 2. Materials. 3. Materials science.
I. Holliday, Douglas.
TA418.9.C6 C66 2022
620.118--dc23

For information on all Willford Press publications
visit our website at www.willfordpress.com

(C WILLFORD PRESS

Contents

Preface

A material made from two or more constituent materials is known as composite material. The physical and chemical properties of the constituent materials are generally significantly different. The characteristics of the resultant material are also different from the source materials. Composites are different from mixtures and solid solutions due to the individual components remaining separate and distinct within the resultant structure. The new material can be lighter, stronger or less expensive compared to the constituent materials. A few examples of engineered composite materials are composite wood, reinforced concrete and metal matrix composites. Composite materials are mostly used in building bridges, buildings, cultured marble sinks and racing car bodies. The extensive content of this book provides the readers with a thorough understanding of composite materials. This book, with its detailed analyses and data, will prove immensely beneficial to professionals and students involved in this area at various levels.

A foreword of all chapters of the book is provided below:

Chapter 1 - The materials which are made from two or more materials that have considerably different chemical or physical properties are known as composite materials. The properties of the resulting material are different from the original materials. All the diverse principles of composite materials as well as their advantages and disadvantages have been briefly introduced in this chapter.; **Chapter 2** - The composite materials which have two or more constituent parts with at least one being a metal are known as metal matrix composites. They are processed through different techniques such as solid state processing, liquid state processing and in-situ processing. The topics elaborated in this chapter will help in gaining a better perspective about these processes as well as the classification of metal matrix composites.; **Chapter 3** - The composite materials which are made up of ceramic fibers which are embedded in a ceramic matrix are known as ceramic matrix composites. Processing of ceramic matrix composites can be done using a variety of techniques such as chemical vapor infiltration and liquid phase infiltration. The diverse aspects of ceramic matrix composites have been thoroughly discussed in this chapter.; **Chapter 4** - The composite material which is composed of a matrix of graphite reinforced by carbon fiber is called carbon-carbon composite. Its processing is done using either gas phase impregnation or liquid phase impregnation process. The topics elaborated in this chapter will help in gaining a better perspective about carbon-carbon composite as well as its processing.; **Chapter 5** - A multiphase solid material in which one of the phases has one, two or three dimensions of less than 100 nanometers is called nanocomposite. There are different types of nanocomposites such as ceramic-matrix nanocomposites, metal-matrix nanocomposites and polymer-matrix nanocomposites. This chapter discusses in detail these types of nanocomposites.

At the end, I would like to thank all the people associated with this book devoting their precious time and providing their valuable contributions to this book. I would also like to express my gratitude to my fellow colleagues who encouraged me throughout the process.

Douglas Holliday

Introduction to Composites Materials

The materials which are made from two or more materials that have considerably different chemical or physical properties are known as composite materials. The properties of the resulting material are different from the original materials. All the diverse principles of composite materials as well as their advantages and disadvantages have been briefly introduced in this chapter.

A composite material is made by combining two or more materials – often ones that have very different properties. The two materials work together to give the composite unique properties. However, within the composite you can easily tell the different materials apart as they do not dissolve or blend into each other.

Many of our modern technologies require materials with unusual combinations of properties that cannot be met by the conventional metal alloys, ceramics, and polymeric materials. This is especially true for materials that are needed for aerospace, underwater, and transportation applications. For example, aircraft engineers are increasingly searching for structural materials that have low densities, are strong, stiff, and abrasion and impact resistant, and are not easily corroded. This is a rather formidable combination of characteristics. Frequently, strong materials are relatively dense; also, increasing the strength or stiffness generally results in a decrease in impact strength.

Material property combinations and ranges have been, and are yet being, extended by the development of composite materials. Generally speaking, a composite is considered to be any multiphase material that exhibits a significant proportion of the properties of both constituent phases such that a better combination of properties is realized. According to this principle of combined action, better property combinations are fashioned by the judicious combination of two or more distinct materials. Property trade-offs are also made for many composites.

Composites of sorts have already been discussed; these include multiphase metal alloys, ceramics, and polymers. For example, pearlitic steels have a microstructure consisting of alternating layers of ferrite and cementite. The ferrite phase is soft and ductile, whereas cementite is hard and very brittle. The combined mechanical characteristics of the pearlite (reasonably high ductility and strength) are superior to those of either of the constituent phases. There are also a number of composites that occur in nature. For example, wood consists of strong and flexible cellulose fibers surrounded and held together by a stiffer material called lignin. Also, bone is a composite of the strong yet soft protein collagen and the hard, brittle mineral apatite.

A composite, in the present context, is a multiphase material that is artificially made, as opposed to one that occurs or forms naturally. In addition, the constituent phases must be chemically dissimilar and separated by a distinct interface. Thus, most metallic alloys and many ceramics do not fit this definition because their multiple phases are formed as a consequence of natural phenomena.

In designing composite materials, scientists and engineers have ingeniously combined various metals, ceramics, and polymers to produce a new generation of extraordinary materials. Most composites have been created to improve combinations of mechanical characteristics such as stiffness, toughness, and ambient and high-temperature strength.

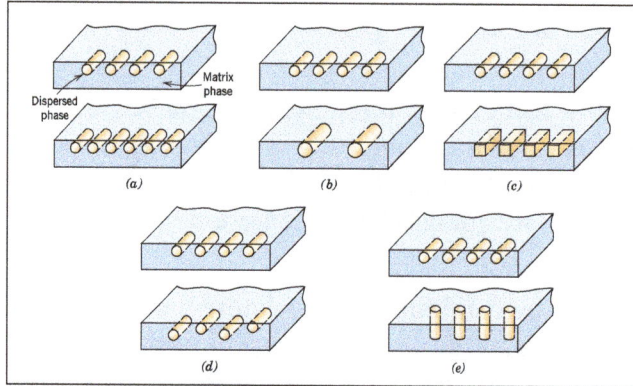

Schematic representations of the various geometrical and spatial characteristics of particles of the dispersed phase that may influence the properties of composites: (a) concentration, (b) size, (c) shape, (d) distribution, and (e) orientation.

Many composite materials are composed of just two phases; one is termed the matrix, which is continuous and surrounds the other phase, often called the dispersed phase. The properties of composites are a function of the properties of the constituent phases, their relative amounts, and the geometry of the dispersed phase. "Dispersed phase geometry" in this context means the shape of the particles and the particle size, distribution, and orientation; these characteristics are represented in figure.

One simple scheme for the classification of composite materials is shown in figure, which consists of three main divisions: particle-reinforced, fiber-reinforced, and structural composites; also, at least two subdivisions exist for each. The dispersed phase for particle-reinforced composites is equiaxed (i.e., particle dimensions are approximately the same in all directions); for fiber-reinforced composites, the dispersed phase has the geometry of a fiber (i.e., a large length-to-diameter ratio). Structural composites are combinations of composites and homogeneous materials.

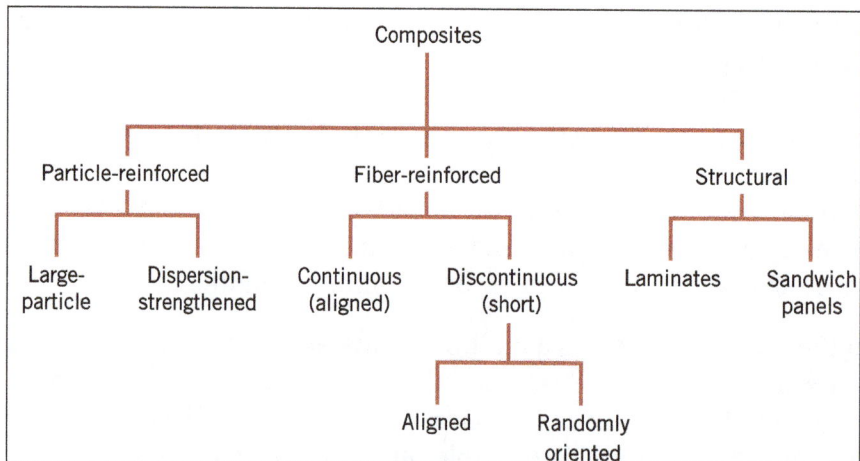

A classification scheme for the various composite types.

Particle-reinforced Composites

As noted in figure large-particle and dispersion-strengthened composites are the two sub-classifications of particle-reinforced composites. The distinction between these is based upon reinforcement or strengthening mechanism. The term "large" is used to indicate that particle–matrix interactions cannot be treated on the atomic or molecular level; rather, continuum mechanics is used. For most of these composites, the particulate phase is harder and stiffer than the matrix. These reinforcing particles tend to restrain movement of the matrix phase in the vicinity of each particle. In essence, the matrix transfers some of the applied stress to the particles, which bear a fraction of the load. The degree of reinforcement or improvement of mechanical behavior depends on strong bonding at the matrix–particle interface.

For dispersion-strengthened composites, particles are normally much smaller, with diameters between 0.01 and 0.1 m (10 and 100 nm). Particle–matrix interactions that lead to strengthening occur on the atomic or molecular level. The mechanism of strengthening is similar to that for precipitation hardening. Whereas the matrix bears the major portion of an applied load, the small dispersed particles hinder or impede the motion of dislocations. Thus, plastic deformation is restricted such that yield and tensile strengths, as well as hardness, improve.

Large-particle Composites

Some polymeric materials to which fillers have been added are really large-particle composites. Again, the fillers modify or improve the properties of the material and replace some of the polymer volume with a less expensive material— the filler.

Another familiar large-particle composite is concrete, which is composed of cement (the matrix), and sand and gravel (the particulates).

Particles can have quite a variety of geometries, but they should be of approximately the same dimension in all directions (equiaxed). For effective reinforcement, the particles should be small and evenly distributed throughout the matrix. Furthermore, the volume fraction of the two phases influences the behavior; mechanical properties are enhanced with increasing particulate content. Two mathematical expressions have been formulated for the dependence of the elastic modulus on the volume fraction of the constituent phases for a two-phase composite. These rules of mixtures equations predict that the elastic modulus should fall between an upper bound represented by:

$$E_c\left(u\right)E_m V_m + E_p V_p$$

and a lower bound, or limit,

$$E_c\left(l\right) = \frac{E_m E_p}{V_m E_p + V_p E_m}$$

In these expressions, E and V denote the elastic modulus and volume fraction, respectively, whereas the subscripts c, m, and p represent composite, matrix, and particulate phases. Figure plots upper-and lower-bound E_c versus-V_p curves for a copper–tungsten composite, in which tungsten is the particulate phase; experimental data points fall between the two curves. Equations analogous to and for fiber-reinforced composites are derived influence of Fiber Orientation and Concentration.

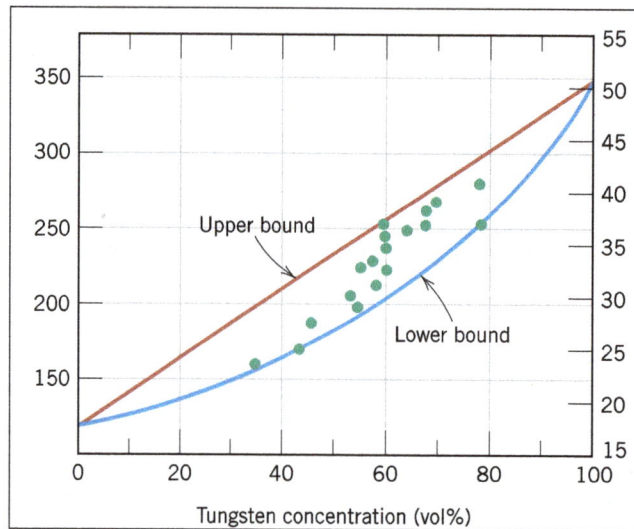

Modulus of elasticity versus volume percent tungsten for a composite of tungsten particles dispersed within a copper matrix.

Large-particle composites are utilized with all three material types (metals, polymers, and ceramics). The cermets are examples of ceramic–metal composites. The most common cermet is the cemented carbide, which is composed of extremely hard particles of a refractory carbide ceramic such as tungsten carbide (WC) or titanium carbide (TiC), embedded in a matrix of a metal such as cobalt or nickel. These composites are utilized extensively as cutting tools for hardened steels. The hard carbide particles provide the cutting surface but, being extremely brittle, are not themselves capable of withstanding the cutting stresses. Toughness is enhanced by their inclusion in the ductile metal matrix, which isolates the carbide particles from one another and prevents particle-to-particle crack propagation. Both matrix and particulate phases are quite refractory, to withstand the high temperatures generated by the cutting action on materials that are extremely hard. No single material could possibly provide the combination of properties possessed by a cermet. Relatively large volume fractions of the particulate phase may be utilized, often exceeding 90 vol%; thus the abrasive action of the composite is maximized. A photomicrograph of a WC–Co cemented carbide is shown in figure.

Both elastomers and plastics are frequently reinforced with various particulate materials. Our use of many of the modern rubbers would be severely restricted without reinforcing particulate materials such as carbon black. Carbon black consists of very small and essentially spherical particles of carbon, produced by the combustion of natural gas or oil in an atmosphere that has only a limited air supply. When added to vulcanize rubber, this extremely inexpensive material enhances tensile strength, toughness, and tear and abrasion resistance. Automobile tires contain on the order of 15 to 30 vol% of carbon black. For the carbon black to provide significant reinforcement, the particle size must be extremely small, with diameters between 20 and 50 nm; also, the particles must be evenly distributed throughout the rubber and must form a strong adhesive bond with the rubber matrix. Particle reinforcement using other materials (e.g., silica) is much less effective because this special interaction between the rubber molecules and particle surfaces does not exist. Figure is an electron micrograph of a carbon black-reinforced rubber.

Photomicrograph of a WC–Co cemented carbide. Light areas are the cobalt matrix; dark regions, the particles of tungsten carbide.

Concrete

Concrete is a common large-particle composite in which both matrix and dispersed phases are ceramic materials. Since the terms "concrete" and "cement" are sometimes incorrectly interchanged, perhaps it is appropriate to make a distinction between them. In a broad sense, concrete implies a composite material consisting of an aggregate of particles that are bound together in a solid body by some type of binding medium, that is a cement. The two most familiar concretes are those made with portland and asphaltic cements, where the aggregate is gravel and sand. Asphaltic concrete is widely used primarily as a paving material, whereas portland cement concrete is employed extensively as a structural building material.

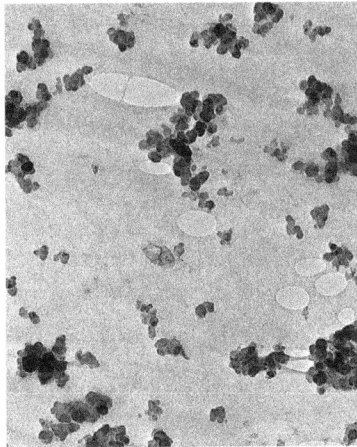

Electron micrograph showing the spherical reinforcing carbon black particles in a synthetic rubber tire tread compound. The areas resembling water marks are tiny air pockets in the rubber.

Portland Cement Concrete

The ingredients for this concrete are portland cement, a fine aggregate (sand), a coarse aggregate (gravel), and water. The aggregate particles act as a filler material to reduce the overall cost of the concrete product because they are cheap, whereas cement is relatively expensive. To achieve the optimum strength and workability of a concrete mixture, the ingredients must be added in the correct proportions. Dense packing of the aggregate and good interfacial contacts are achieved by having particles of two different sizes; the fine particles of sand should fill the void spaces between the gravel particles. Ordinarily these aggregates comprise between 60% and 80% of the total volume. The amount of cement–water paste should be sufficient to coat all the sand and gravel particles, otherwise the cementitious bond will be incomplete. Furthermore, all the constituents should be thoroughly mixed. Complete bonding between cement and the aggregate particles is contingent upon the addition of the correct quantity of water. Too little water leads to incomplete bonding, and too much results in excessive porosity; in either case the final strength is less than the optimum.

The character of the aggregate particles is an important consideration. In particular, the size distribution of the aggregates influences the amount of cement–water paste required. Also, the surfaces should be clean and free from clay and silt, which prevent the formation of a sound bond at the particle surface.

Portland cement concrete is a major material of construction, primarily because it can be poured in place and hardens at room temperature, and even when submerged in water. However, as a structural material, there are some limitations and disadvantages. Like most ceramics, portland cement concrete is relatively weak and extremely brittle; its tensile strength is approximately 10 to 15 times smaller than its compressive strength. Also, large concrete structures can experience considerable thermal expansion and contraction with temperature fluctuations. In addition, water penetrates into external pores, which can cause severe cracking in cold weather as a consequence of freeze–thaw cycles. Most of these inadequacies may be eliminated or at least improved by reinforcement and the incorporation of additives.

Reinforced Concrete

The strength of portland cement concrete may be increased by additional reinforcement. This is usually accomplished by means of steel rods, wires, bars (rebar), or mesh, which are embedded into the fresh and uncured concrete. Thus, the reinforcement renders the hardened structure capable of supporting greater tensile, compressive, and shear stresses. Even if cracks develop in the concrete, considerable reinforcement is maintained.

Steel serves as a suitable reinforcement material because its coefficient of thermal expansion is nearly the same as that of concrete. In addition, steel is not rapidly corroded in the cement environment, and a relatively strong adhesive bond is formed between it and the cured concrete. This adhesion may be enhanced by the incorporation of contours into the surface of the steel member, which permits a greater degree of mechanical interlocking.

Portland cement concrete may also be reinforced by mixing into the fresh concrete fibers of a high-modulus material such as glass, steel, nylon, and polyethylene. Care must be exercised in utilizing this type of reinforcement, since some fiber materials experience rapid deterioration when exposed to the cement environment.

Still another reinforcement technique for strengthening concrete involves the introduction of residual compressive stresses into the structural member; the resulting material is called prestressed concrete. This method utilizes one characteristic of brittle ceramics—namely, that they are stronger in compression than in tension. Thus, to fracture a prestressed concrete member, the magnitude of the precompressive stress must be exceeded by an applied tensile stress.

In one such prestressing technique, high-strength steel wires are positioned inside the empty molds and stretched with a high tensile force, which is maintained constant. After the concrete has been placed and allowed to harden, the tension is released. As the wires contract, they put the structure in a state of compression because the stress is transmitted to the concrete via the concrete–wire bond that is formed.

Another technique is also utilized in which stresses are applied after the concrete hardens; it is appropriately called posttensioning. Sheet metal or rubber tubes are situated inside and pass through the concrete forms, around which the concrete is cast. After the cement has hardened, steel wires are fed through the resulting holes, and tension is applied to the wires by means of jacks attached and abutted to the faces of the structure. Again, a compressive stress is imposed on the concrete piece, this time by the jacks. Finally, the empty spaces inside the tubing are filled with a grout to protect the wire from corrosion.

Concrete that is prestressed should be of a high quality, with a low shrinkage and a low creep rate. Prestressed concretes, usually prefabricated, are commonly used for highway and railway bridges.

Dispersion-strengthened Composites

Metals and metal alloys may be strengthened and hardened by the uniform dispersion of several volume percent of fine particles of a very hard and inert material. The dispersed phase may be metallic or nonmetallic; oxide materials are often used. Again, the strengthening mechanism involves interactions between the particles and dislocations within the matrix, as with precipitation hardening. The dispersion strengthening effect is not as pronounced as with precipitation hardening; however, the strengthening is retained at elevated temperatures and for extended time periods because the dispersed particles are chosen to be unreactive with the matrix phase. For precipitation-hardened alloys, the increase in strength may disappear upon heat treatment as a consequence of precipitate growth or dissolution of the precipitate phase.

The high-temperature strength of nickel alloys may be enhanced significantly by the addition of about 3 vol% of thoria (ThO_2) as finely dispersed particles; this material is known as thoria-dispersed (or TD) nickel. The same effect is produced in the aluminum–aluminum oxide system. A very thin and adherent alumina coating is caused to form on the surface of extremely small (0.1 to 0.2 m thick) flakes of aluminum, which are dispersed within an aluminum metal matrix; this material is termed sintered aluminum powder (SAP).

Fiber-reinforced Composites

Technologically, the most important composites are those in which the dispersed phase is in the form of a fiber. Design goals of fiber-reinforced composites often include high strength and/or stiffness on a weight basis. These characteristics are expressed in terms of specific strength and specific modulus parameters, which correspond, respectively, to the ratios of tensile strength to specific gravity and modulus of elasticity to specific gravity. Fiber-reinforced composites with exceptionally high specific strengths and moduli have been produced that utilize low-density fiber and matrix materials.

Fiber-reinforced composites are sub-classified by fiber length. For short fiber, the fibers are too short to produce a significant improvement in strength.

Influence of Fiber Length

The mechanical characteristics of a fiber-reinforced composite depend not only on the properties of the fiber, but also on the degree to which an applied load is transmitted to the fibers by the matrix phase. Important to the extent of this load transmittance is the magnitude of the interfacial bond between the fiber and matrix phases. Under an applied stress, this fiber–matrix bond ceases at the fiber ends, yielding a matrix deformation pattern as shown schematically in figure; in other words, there is no load transmittance from the matrix at each fiber extremity.

Some critical fiber length is necessary for effective strengthening and stiffening of the composite material. This critical length l_c is dependent on the fiber diameter d and its ultimate (or tensile) strength σ_f^* and on the fiber–matrix bond strength (or the shear yield strength of the matrix, whichever is smaller) τ_c according to,

$$l_c = \frac{\sigma_f^* d}{2\tau_c}$$

For a number of glass and carbon fiber–matrix combinations, this critical length is on the order of 1 mm, which ranges between 20 and 150 times the fiber diameter.

When a stress equal to σ_f^* is applied to a fiber having just this critical length, the stress–position profile shown in figure results; that is, the maximum fiber load is achieved only at the axial center of the fiber. As fiber length l increases, the fiber reinforcement becomes more effective; this is demonstrated in figure, a stress–axial position profile for $l > l_c$ when the applied stress is equal to the fiber strength. Figure shows the stress–position profile for $l < l_c$.

Fibers for which $l \gg l_c$ (normally $l > 15 l_c$) are termed continuous; discontinuousor short fiber shave lengths shorter than this. For discontinuous fibers of lengths significantly less than the matrix deforms around the fiber such that there is virtually no stress transference and little reinforcement by the fiber. To affect a significant improvement in strength of the composite, the fibers must be continuous.

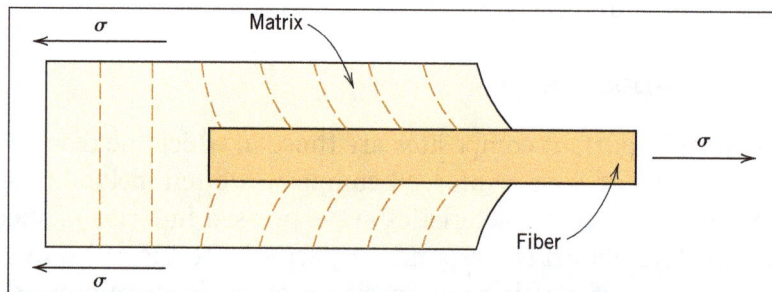

The deformation pattern in the matrix surrounding a fiber that is subjected to an applied tensile load.

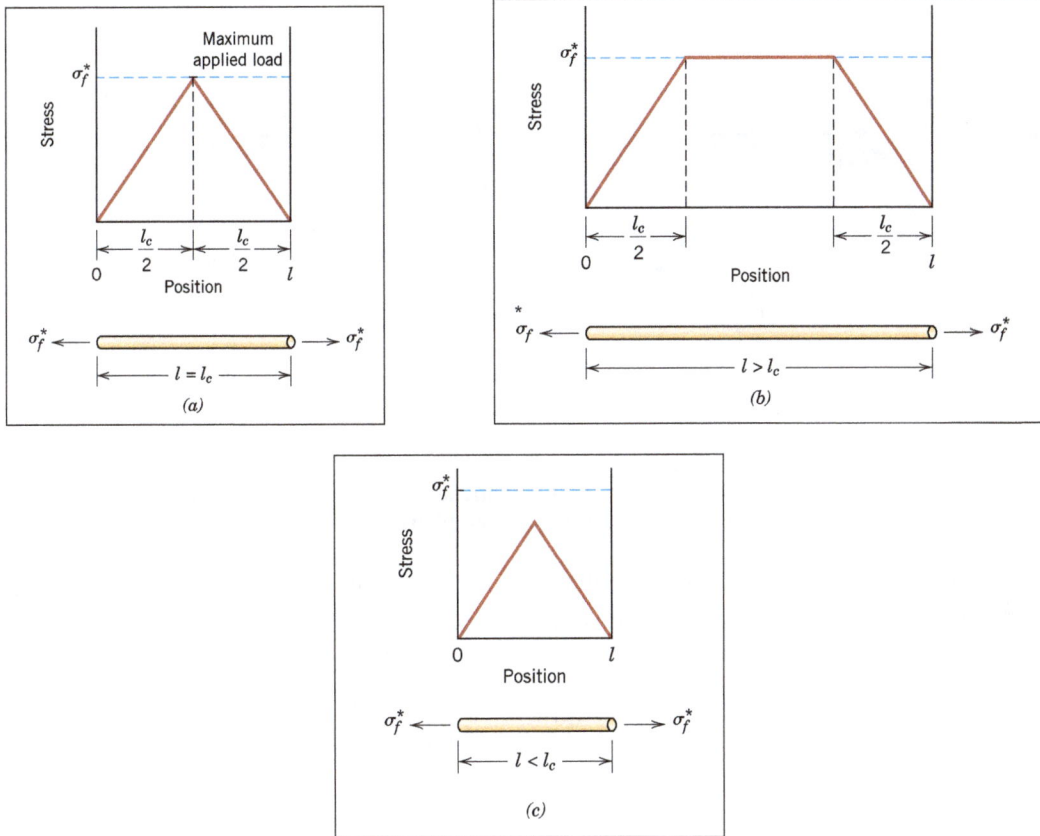

Stress–position profiles when fiber length l (a) is equal to the critical length lc (b) is greater than the critical length, and (c) is less than the critical length for a fiber-reinforced composite that is subjected to a tensile stress equal to the fiber tensile strength σ_f^*.

Influence of Fiber Orientation and Concentration

The arrangement or orientation of the fibers relative to one another, the fiber concentration, and the distribution all have a significant influence on the strength and other properties of fiber-reinforced composites. With respect to orientation, two extremes are possible: (1) a parallel alignment of the longitudinal axis of the fibers in a single direction, and (2) a totally random alignment. Continuous fibers are normally aligned, whereas discontinuous fibers may be aligned, randomly oriented or partially oriented. Better overall composite properties are realized when the fiber distribution is uniform.

Continuous and Aligned Fiber Composites

Tensile Stress–strain Behavior

Mechanical responses of this type of composite depend on several factors to include the stress–strain behaviors of fiber and matrix phases, the phase volume fractions, and, in addition, the direction in which the stress or load is applied. Furthermore, the properties of a composite having its fibers aligned are highly anisotropic, that is, dependent on the direction in which they are measured. Let us first consider the stress–strain behavior for the situation wherein the stress is applied along the direction of alignment, the longitudinal direction, which is indicated in figure.

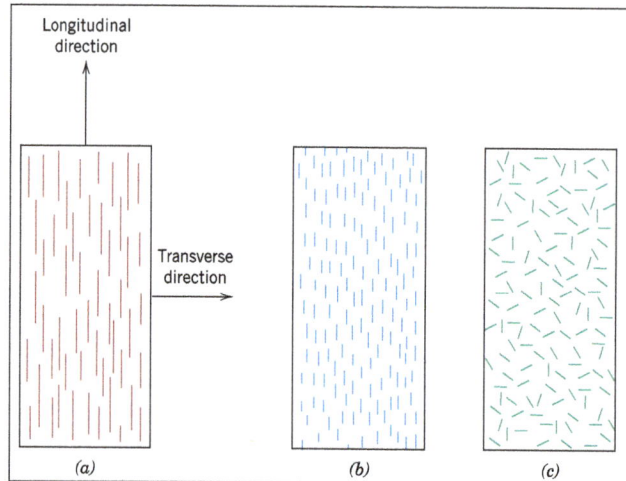

Schematic representations of (a) continuous and aligned, (b) discontinuous and aligned, and (c) discontinuous and randomly oriented fiberreinforced composites.

To begin, assume the stress versus strain behaviors for fiber and matrix phases that are represented schematically in figure in this treatment we consider the fiber to be totally brittle and the matrix phase to be reasonably ductile. Also indicated in this figure are fracture strengths in tension for fiber and matrix, σ_f^* and σ_m^* respectively, and their corresponding fracture strains, ϵ_f^* and ϵ_m^* furthermore, it is assumed that $\epsilon_m^* > \epsilon_f^*$ which is normally the case.

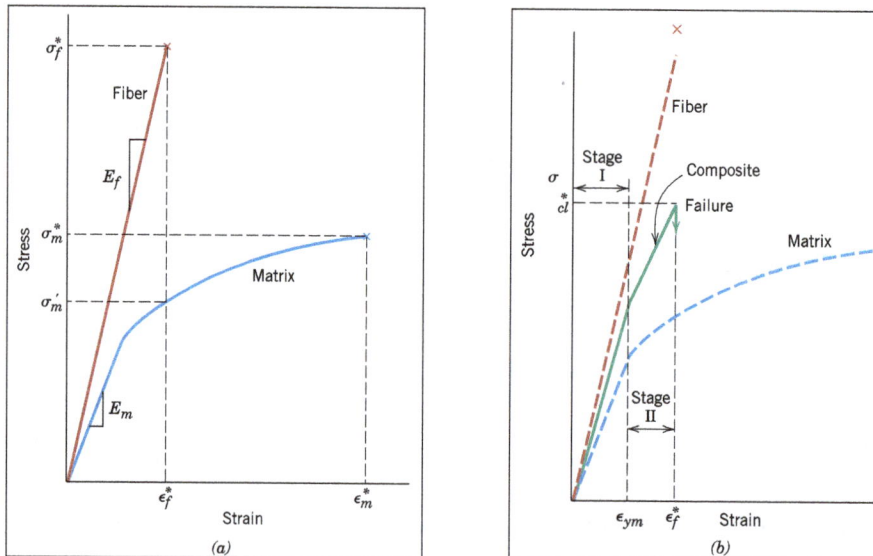

In figure represents (a) Schematic stress–strain curves for brittle fiber and ductile matrix materials. Fracture stresses and strains for both materials are noted. (b) Schematic stress–strain curve for an aligned fiber-reinforced composite that is exposed to a uniaxial stress applied in the direction of alignment; curves for the fiber and matrix materials shown in part (a) are also superimposed.

A fiber-reinforced composite consisting of these fiber and matrix materials will exhibit the uniaxial stress–strain response illustrated in figure the fiber and matrix behaviors from figure are included to provide perspective. In the initial Stage I region, both fibers and matrix deform elastically; normally this portion of the curve is linear. Typically, for a composite of this type, the matrix yields and deforms plastically while the fibers continue to stretch elastically, inasmuch as the tensile strength of the fibers is significantly higher than the yield strength of the matrix. This process constitutes Stage II as noted in the figure; this stage is ordinarily very nearly linear, but of diminished slope relative to Stage I. Furthermore, in passing from Stage I to Stage II, the proportion of the applied load that is borne by the fibers increases.

The onset of composite failure begins as the fibers start to fracture, which corresponds to a strain of approximately as noted in figure Composite failure is not catastrophic for a couple of reasons. First, not all fibers fracture at the same time, since there will always be considerable variations in the fracture strength of brittle fiber materials. In addition, even after fiber failure, the matrix is still intact inasmuch as $\epsilon_f^* < \epsilon_m^*$. Thus, these fractured fibers, which are shorter than the original ones, are still embedded within the intact matrix, and consequently are capable of sustaining a diminished load as the matrix continues to plastically deform.

Elastic Behavior—Longitudinal Loading

Let us now consider the elastic behavior of a continuous and oriented fibrous composite that is loaded in the direction of fiber alignment. First, it is assumed that the fiber–matrix interfacial bond is very good, such that deformation of both matrix and fibers is the same (an isostrain situation). Under these conditions, the total load sustained by the composite F_c is equal to the sum of the loads carried by the matrix phase F_m and the fiber phase F_f or:

$$F_c = F_m + F_f$$

From the definition of stress, Equation $E_c(u) E_m V_m + E_p V_p$, $F = \sigma A$, and thus expressions for F_c, F_m and F_f in terms of their respective stresses (σ_c, σ_m and σ_f) and cross-sectional areas (A_c, A_m and A_f) are possible. Substitution of these into Equation $F_c = F_m + F_f$ yields:

$$\sigma_c A_c = \sigma_m A_m + \sigma_f A_f$$

and then, dividing through by the total cross-sectional area of the composite, A_c we have:

$$\sigma_c = \sigma_m \frac{A_m}{A_c} + \sigma_f \frac{A_f}{A_c}$$

where A_m / A_c and A_f / A_c are the area fractions of the matrix and fiber phases, respectively. If the composite, matrix, and fiber phase lengths are all equal, A_m / A_c is equivalent to the volume fraction of the matrix, and likewise for the fibers, $V_f = A_f / A_c$.

Equation $\sigma_c = \sigma_m \dfrac{A_m}{A_c} + \sigma_f \dfrac{A_f}{A_c}$ now becomes,

$$\sigma_c = \sigma_m V_m + \sigma_f V_f$$

The previous assumption of an isostrain state means that,

$$\epsilon_c = \epsilon_m = \epsilon_f$$

and when each term in Equation $\sigma_c = \sigma_m V_m + \sigma_f V_f$ is divided by its respective strain,

$$\frac{\sigma_c}{\epsilon_c} = \frac{\sigma_m}{\epsilon_m} V_m + \frac{\sigma_f}{\epsilon_f} V_f.$$

Furthermore, if composite, matrix, and fiber deformations are all elastic, then $\sigma_c / \epsilon_c = E_c, \sigma_m / \epsilon_m = E_m$ and $\sigma_f / \epsilon_f = E_f$ the E's being the moduli of elasticity for the respective phases. Substitution into

Equation $\frac{\sigma_c}{\epsilon_c} = \frac{\sigma_m}{\epsilon_m} V_m + \frac{\sigma_f}{\epsilon_f} V_f$ yields an expression for the modulus of elasticity of a continuous and

aligned fibrous composite in the direction of alignment (or longitudinal direction), E_{cl} as,

$$E_{cl} = E_m V_m + E_f V_f$$

$$E_{cl} = E_m \left(1 - V_f\right) + E_f V_f$$

since the composite consists of only matrix and fiber phases; that is, $V_m + V_f = 1$.

Thus, E_{cl} is equal to the volume-fraction weighted average of the moduli of elasticity of the fiber and matrix phases. Other properties, including density, also have this dependence on volume fractions. Equation $E_{cl} = E_m V_m + E_f V_f$ is the fiber analogue of Equation $E_c(u) E_m V_m + E_p V_p$, the upper bound for particle-reinforced composites.

It can also be shown, for longitudinal loading that the ratio of the load carried by the fibers to that carried by the matrix is,

$$\frac{F_f}{F_m} = \frac{E_f V_f}{E_m V_m}$$

The demonstration is left as a homework problem.

Elastic Behavior—Transverse Loading

A continuous and oriented fiber composite may be loaded in the transverse direction; that is, the load is applied at an angle to the direction of fiber alignment as shown in figure. For this situation the stress to which the composite as well as both phases are exposed is the same, or,

$$\sigma_c = \sigma_m + \sigma_f = \sigma$$

This is termed an isostress state. Also, the strain or deformation of the entire composite ϵ_c is,

$$\epsilon_m = \epsilon_m V_m + \epsilon_f V_f$$

but, since $\epsilon = \sigma / E$.

$$\frac{\sigma}{E_{ct}} = \frac{\sigma}{E_m} V_m + \frac{\sigma}{E_f} V_f$$

where E_{cl} is the modulus of elasticity in the transverse direction. Now, dividing through by σ yields,

$$\frac{1}{E_{ct}} = \frac{V_m}{E_m} V_m + \frac{\sigma}{E_f} V_f$$

which reduces to,

$$E_{ct} = \frac{E_m E_f}{V_m E_f + V_f E_m} = \frac{E_m E_f}{\left(1 - V_f\right) E_f + V_f E_m}$$

Equation $E_{ct} = \dfrac{E_m E_f}{V_m E_f + V_f E_m} = \dfrac{E_m E_f}{\left(1 - V_f\right) E_f + V_f E_m}$ is analogous to the lower-bound expression

for particulate composites, equation $E_c\left(l\right) = \dfrac{E_m E_p}{V_m E_p + V_p E_m}$.

Table: Typical longitudinal and transverse tensile strengths for three unidirectional fiber-reinforced composites. the fiber content for each is approximately 50 vol%.

Material	Longitudinal Tensile Strength (MPa)	Transverse Tensile Strength (MPa)
Glass–polyester	700	20
Carbon (high modulus)–epoxy	1000	35
Kevlar™–epoxy	1200	20

Longitudinal Tensile Strength

We now consider the strength characteristics of continuous and aligned fiber-reinforced composites that are loaded in the longitudinal direction. Under these circumstances, strength is normally taken as the maximum stress on the stress–strain curve, figure often this point corresponds to fiber fracture, and marks the onset of composite failure. Table lists typical longitudinal tensile strength values for three common fibrous composites. Failure of this type of composite material is a relatively complex process, and several different failure modes are possible. The mode that operates for a specific composite will depend on fiber and matrix properties, and the nature and strength of the fiber–matrix interfacial bond.

If we assume that $\epsilon_f^* < \epsilon_m^*$ which is the usual case, then fibers will fail before the matrix. Once the fibers have fractured, the majority of the load that was borne by the fibers is now transferred to the matrix. This being the case, it is possible to adapt the expression for the stress on this type of composite, Equation $\sigma_c = \sigma_m V_m + \sigma_f V_f$, into the following expression for the longitudinal strength of the composite σ_{cl}^*:

$$\sigma_{cl}^* \quad \sigma_m' \left(1 - V_f\right) + * \sigma_f^* V_f$$

Here σ_m' is the stress in the matrix at fiber failure and, as previously, σ_f^* is the fiber tensile strength.

Transverse Tensile Strength

The strengths of continuous and unidirectional fibrous composites are highly anisotropic, and such composites are normally designed to be loaded along the high strength, longitudinal direction. However, during in-service applications transverse tensile loads may also be present. Under these circumstances, premature failure may result inasmuch as transverse strength is usually extremely low—it sometimes lies below the tensile strength of the matrix. Thus, in actual fact, the reinforcing effect of the fibers is a negative one. Typical transverse tensile strengths for three unidirectional composites are contained in table.

Whereas longitudinal strength is dominated by fiber strength, a variety of factors will have a significant influence on the transverse strength; these factors include properties of both the fiber and matrix, the fiber–matrix bond strength, and the presence of voids. Measures that have been employed to improve the transverse strength of these composites usually involve modifying properties of the matrix.

Discontinuous and Aligned Fiber Composites

Even though reinforcement efficiency is lower for discontinuous than for continuous fibers discontinuous and aligned fiber composites are becoming increasingly more important in the commercial market. Chopped glass fibers are used most extensively; however, carbon and aramid discontinuous fibers are also employed. These short fiber composites can be produced having moduli of elasticity and tensile strengths that approach 90% and 50%, respectively, of their continuous fiber counterparts.

For a discontinuous and aligned fiber composite having a uniform distribution of fibers and in which $l < l_c$ the longitudinal strength σ_{cd}^* is given by the relationship,

$$\sigma_{cd}^* = \sigma_f^* V_f \left(1 - \frac{l_c}{2l}\right) + \sigma_m' \left(1 - V_f\right)$$

where σ_f^* and σ_m' represent, respectively, the fracture strength of the fiber and the stress in the matrix when the composite fails.

If the fiber length is less than critical $(l < l_c)$ then the longitudinal strength $\left(\sigma_{cd}^*\right)$ is given by,

$$\sigma_{cd}^* = \frac{l \tau_c}{d} V_f + \sigma_m' \left(1 - V_f\right)$$

where d is the fiber diameter and τ_c is the smaller of either the fiber–matrix bond strength or the matrix shear yield strength.

Discontinuous and Randomly Oriented Fiber Composites

Normally, when the fiber orientation is random, short and discontinuous fibers are used; reinforcement of this type is schematically demonstrated in figure. Under these circumstances, a "rule-of-mixtures" expression for the elastic modulus similar to equation $E_{cl} = E_m V_m + E_f V_f$ may be utilized, as follows:

$$E_{cd} = K E_f V_f + E_m V_m$$

Table: Properties of unreinforced and reinforced polycarbonates with randomly oriented glass fibers.

Property	Fiber Reinforcement (vol%)			
	Unreinforced	20	30	40
Specific gravity	1.19–1.22	1.35	1.43	1.52
Tensile strength	59–62	110	131	159
[MPa (ksi)]	(8.5–9.0)	(16)	(19)	(23)
Modulus of elasticity	2.24–2.345	5.93	8.62	11.6
[GPa (10^6 psi)]	(0.325–0.340)	(0.86)	(1.25)	(1.68)
Elongation (%)	90–115	4–6	3–5	3–5
Impact strength,	12–16	2.0	2.0	2.5

In this expression, K is a fiber efficiency parameter that depends on V_f and the E_f / E_m ratio. Of course, its magnitude will be less than unity, usually in the range 0.1 to 0.6. Thus, for random fiber reinforcement (as with oriented), the modulus increases in some proportion of the volume fraction of fiber. Table which gives some of the mechanical properties of unreinforced and reinforced polycarbonates for discontinuous and randomly oriented glass fibers provides an idea of the magnitude of the reinforcement that is possible.

Aligned fibrous composites are inherently anisotropic, in that the maximum strength and reinforcement are achieved along the alignment (longitudinal) direction. In the transverse direction, fiber reinforcement is virtually nonexistent: fracture usually occurs at relatively low tensile stresses. For other stress orientations, composite strength lies between these extremes. The efficiency of fiber reinforcement for several situations is presented in table this efficiency is taken to be unity for an oriented fiber composite in the alignment direction, and zero perpendicular to it.

When multidirectional stresses are imposed within a single plane, aligned layers that are fastened together one on top of another at different orientations are frequently utilized. These are termed laminar composites.

Table: Reinforcement efficiency of fiber-reinforced composites for several fiber orientations and at various directions of stress application.

Fiber Orientation	Stress Direction	Reinforcement Efficiency
All fibers parallel	Parallel to fibers	$\frac{1}{}$
	Perpendicular to fibers	0
Fibers randomly and uniformly distributed within a specific plane	Any direction in the plane of the fibers	$\frac{3}{8}$
Fibers randomly and uniformly distributed within three dimensions in space	Any direction	$\frac{1}{5}$

Applications involving totally multidirectional applied stresses normally use discontinuous fibers, which are randomly oriented in the matrix material. Table shows that the reinforcement efficiency is only one-fifth that of an aligned composite in the longitudinal direction; however, the mechanical characteristics are isotropic.

Consideration of orientation and fiber length for a particular composite will depend on the level and nature of the applied stress as well as fabrication cost. Production rates for short-fiber composites (both aligned and randomly oriented) are rapid, and intricate shapes can be formed that are not possible with continuous fiber reinforcement. Furthermore, fabrication costs are considerably lower than for continuous and aligned; fabrication techniques applied to short-fiber composite materials include compression, injection, and extrusion molding.

Fiber Phase

An important characteristic of most materials, especially brittle ones, is that a small diameter fiber is much stronger than the bulk material. The probability of the presence of a critical surface flaw that can lead to fracture diminishes with decreasing specimen volume, and this feature is used to advantage in the fiber-reinforced composites. Also, the materials used for reinforcing fibers have high tensile strengths.

On the basis of diameter and character, fibers are grouped into three different classifications: whiskers, fibers, and wires. Whiskers are very thin single crystals that have extremely large length-to-diameter ratios. As a consequence of their small size, they have a high degree of crystalline perfection and are virtually flaw free, which accounts for their exceptionally high strengths; they are among the strongest known materials. In spite of these high strengths, whiskers are not utilized extensively as a reinforcement medium because they are extremely expensive. Moreover, it is difficult and often impractical to incorporate whiskers into a matrix. Whisker materials include graphite, silicon carbide, silicon nitride, and aluminum oxide; some mechanical characteristics of these materials are given in table.

Materials that are classified as fibers are either polycrystalline or amorphous and have small diameters; fibrous materials are generally either polymers or ceramics (e.g., the polymer aramids, glass, carbon, boron, aluminum oxide, and silicon carbide). Table also presents some data on a few materials that are used in fiber form.

Fine wires have relatively large diameters; typical materials include steel, molybdenum, and tungsten. Wires are utilized as a radial steel reinforcement in automobile tires, in filament-wound rocket casings, and in wire-wound high-pressure hoses.

Table: Characteristics of several fiber-reinforcement materials.

Material	Specific Gravity	Tensile Strength [GPa (10^6 psi)]	Specific Strength (GPa)	Modulus of Elasticity [GPa (10^6 psi)]	Specific Modulus (GPa)
		Whiskers			
Graphite	2.2	20	9.1	700	318
		(3)		(100)	
Silicon nitride	3.2	5–7	1.56–2.2	350–380	109–118
		(0.75–1.0)		(50–55)	
Aluminum oxide	4.0	10–20	2.5–5.0	700–1500	175–375
		(1–3)		(100–220)	
Silicon carbide	3.2	20	6.25	480	150
		(3)		(70)	
		Fibers			
Aluminum oxide	3.95	1.38	0.35	379	96
		(0.2)		(55)	
Aramid (Kevlar 49™)	1.44	3.6–4.1	2.5–2.85	131	91
		(0.525–0.600)		(19)	
Carbon[a]	1.78–2.15	1.5–4.8	0.70–2.70	228–724	106–407
		(0.22–0.70)		(32–100)	
E-glass	2.58	3.45	1.34	72.5	28.1
		(0.5)		(10.5)	
Boron	2.57	3.6	1.40	400	156
		(0.52)		(60)	
Silicon carbide	3.0	3.9	1.30	400	133
		(0.57)		(60)	
UHMWPE (Spectra 900™)	0.97	2.6	2.68	117	121
		(0.38)		(17)	
		Metallic Wires			
High-strength steel	7.9	2.39	0.30	210	26.6
		(0.35)		(30)	
Molybdenum	10.2	2.2	0.22	324	31.8
		(0.32)		(47)	
Tungsten	19.3	2.89	0.15	407	21.1
		(0.42)		(59)	

The term "carbon" instead of "graphite" is used to denote these fibers, since they are composed of crystalline graphite regions, and also of noncrystalline material and areas of crystal misalignment.

Matrix Phase

The matrix phase of fibrous composites may be a metal, polymer, or ceramic. In general, metals and polymers are used as matrix materials because some ductility is desirable; for ceramic-matrix composites, the reinforcing component is added to improve fracture toughness.

For fiber-reinforced composites, the matrix phase serves several functions. First, it binds the fibers together and acts as the medium by which an externally applied stress is transmitted and distributed to the fibers; only a very small proportion of an applied load is sustained by the matrix phase. Furthermore, the matrix material should be ductile. In addition, the elastic modulus of the fiber should be much higher than that of the matrix. The second function of the matrix is to protect the individual fibers from surface damage as a result of mechanical abrasion or chemical reactions with the environment. Such interactions may introduce surface flaws capable of forming cracks, which may lead to failure at low tensile stress levels. Finally, the matrix separates the fibers and, by virtue of its relative softness and plasticity, prevents the propagation of brittle cracks from fiber to fiber, which could result in catastrophic failure; in other words, the matrix phase serves as a barrier to crack propagation. Even though some of the individual fibers fail, total composite fracture will not occur until large numbers of adjacent fibers, once having failed, form a cluster of critical size.

It is essential that adhesive bonding forces between fiber and matrix be high to minimize fiber pull-out. In fact, bonding strength is an important consideration in the choice of the matrix–fiber combination. The ultimate strength of the composite depends to a large degree on the magnitude of this bond; adequate bonding is essential to maximize the stress transmittance from the weak matrix to the strong fibers.

Polymer-matrix Composites

Polymer-matrix composites (PMCs) consist of a polymer resin1 as the matrix, with fibers as the reinforcement medium. These materials are used in the greatest diversity of composite applications, as well as in the largest quantities, in light of their room-temperature properties, ease of fabrication, and cost. The various classifications of PMCs are discussed according to reinforcement type (i.e., glass, carbon, and aramid), along with their applications and the various polymer resins that are employed.

Glass Fiber-reinforced Polymer (GFRP) Composites

Fiberglass is simply a composite consisting of glass fibers, either continuous or discontinuous, contained within a polymer matrix; this type of composite is produced in the largest quantities. The composition of the glass that is most commonly drawn into fibers (sometimes referred to as E-glass) is contained in table fiber diameters normally range between 3 and 20 m. Glass is popular as a fiber reinforcement material for several reasons:

1. It is easily drawn into high-strength fibers from the molten state.

2. It is readily available and may be fabricated into a glass-reinforced plastic economically using a wide variety of composite-manufacturing techniques.

3. As a fiber it is relatively strong, and when embedded in a plastic matrix, it produces a composite having a very high specific strength.

4. When coupled with the various plastics, it possesses a chemical inertness that renders the composite useful in a variety of corrosive environments.

The surface characteristics of glass fibers are extremely important because even minute surface flaws can deleteriously affect the tensile properties. Surface flaws are easily introduced by rubbing or abrading the surface with another hard material. Also, glass surfaces that have been exposed to the normal atmosphere for even short time periods generally have a weakened surface layer that interferes with bonding to the matrix. Newly drawn fibers are normally coated during drawing with a "size," a thin layer of a substance that protects the fiber surface from damage and undesirable environmental interactions. This size is ordinarily removed prior to composite fabrication and replaced with a "coupling agent" or finish that produces a chemical bond between the fiber and matrix.

There are several limitations to this group of materials. In spite of having high strengths, they are not very stiff and do not display the rigidity that is necessary for some applications (e.g., as structural members for airplanes and bridges). Most fiberglass materials are limited to service temperatures below at higher temperatures most polymers begin to flow or to deteriorate. Service temperatures may be extended to approximately by using high-purity fused silica for the fibers and high-temperature polymers such as the polyimide resins.

Many fiberglass applications are familiar: automotive and marine bodies, plastic pipes, storage containers, and industrial floorings. The transportation industries are utilizing increasing amounts of glass fiber-reinforced plastics in an effort to decrease vehicle weight and boost fuel efficiencies. A host of new applications are being used or currently investigated by the automotive industry.

Carbon Fiber-reinforced Polymer (CFRP) Composites

Carbon is a high-performance fiber material that is the most commonly used reinforcement in advanced (i.e., non-fiberglass) polymer-matrix composites. The reasons for this are as follows:

1. Carbon fibers have the highest specific modulus and specific strength of all reinforcing fiber materials.

2. They retain their high tensile modulus and high strength at elevated temperatures; high-temperature oxidation, however, may be a problem.

3. At room temperature, carbon fibers are not affected by moisture or a wide variety of solvents, acids, and bases.

4. These fibers exhibit a diversity of physical and mechanical characteristics, allowing composites incorporating these fibers to have specific engineered properties.

5. Fiber and composite manufacturing processes have been developed that are relatively inexpensive and cost effective.

Use of the term "carbon fiber" may seem perplexing since carbon is an element, and, the stable form of crystalline carbon at ambient conditions is graphite, having the structure represented in figure Carbon fibers are not totally crystalline, but are composed of both graphitic and noncrystalline regions; these areas of noncrystallinity are devoid of the three-dimensional ordered arrangement of hexagonal carbon networks that is characteristic of graphite.

Manufacturing techniques for producing carbon fibers are relatively complex and will not be discussed. However, three different organic precursor materials are used: rayon, polyacrylonitrile (PAN), and pitch. Processing technique will vary from precursor to precursor, as will also the resultant fiber characteristics.

One classification scheme for carbon fibers is by tensile modulus; on this basis the four classes are standard, intermediate, high, and ultrahigh moduli. Furthermore, fiber diameters normally range between 4 and 10 mm; both continuous and chopped forms are available. In addition, carbon fibers are normally coated with a protective epoxy size that also improves adhesion with the polymer matrix.

Carbon-reinforced polymer composites are currently being utilized extensively in sports and recreational equipment (fishing rods, golf clubs), filament-wound rocket motor cases, pressure vessels, and aircraft structural components—both military and commercial, fixed wing and helicopters (e.g., as wing, body, stabilizer, and rudder components).

Aramid Fiber-reinforced Polymer Composites

Aramid fibers are high-strength, high-modulus materials that were introduced in the early 1970s. They are especially desirable for their outstanding strength-toweight ratios, which are superior to metals. Chemically, this group of materials is known as poly(paraphenylene terephthalamide). There are a number of aramid materials; trade names for two of the most common are Kevlar™ and Nomex. For the former, there are several grades that have different mechanical behaviors. During synthesis, the rigid molecules are aligned in the direction of the fiber axis, as liquid crystal domains the repeat unit and the mode of chain alignment are represented in figure Mechanically, these fibers have longitudinal tensile strengths and tensile moduli that are higher than other polymeric fiber materials; however, they are relatively weak in compression. In addition, this material is known for its toughness, impact resistance, and resistance to creep and fatigue failure. Even though the aramids are thermoplastics, they are, nevertheless, resistant to combustion and stable to relatively high temperatures; the temperature range over which they retain their high mechanical properties is between 200 and 200 °C (330 and 390 °F). Chemically, they are susceptible to degradation by strong acids and bases, but they are relatively inert in other solvents and chemicals.

The aramid fibers are most often used in composites having polymer matrices; common matrix materials are the epoxies and polyesters. Since the fibers are relatively flexible and somewhat ductile, they may be processed by most common textile operations. Typical applications of these aramid composites are in ballistic products (bulletproof vests and armor), sporting goods, tires, ropes, missile cases, pressure vessels, and as a replacement for asbestos in automotive brake and clutch linings, and gaskets.

The properties of continuous and aligned glass-, carbon-, and aramid-fiber reinforced epoxy composites are included in table. Thus, a comparison of the mechanical characteristics of these three materials may be made in both longitudinal and transverse directions.

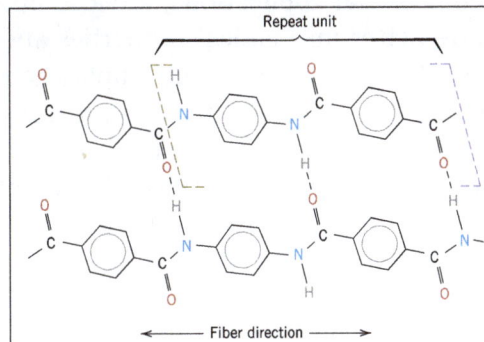

Schematic representation of repeat unit and chain structures for aramid (Kevlar) fibers. Chain alignment with the fiber direction and hydrogen bonds that form between adjacent chains are also shown.

Table: Properties of continuous and aligned glass-, carbon-, and aramid-fiber reinforced epoxy-matrix composites in longitudinal and transverse directions. in all cases the fiber volume fraction is 0.60.

Property	Glass (E-glass)	Carbon (High Strength)	Aramid (Kevlar 49)
Specific gravity	2.1	1.6	1.4
Tensile modulus			
Longitudinal [GPa (10^6 psi)]	45 (6.5)	145 (21)	76 (11)
Transverse [GPa (10^6 psi)]	12 (1.8)	10 (1.5)	5.5 (0.8)
Tensile strength			
Longitudinal [MPa (ksi)]	1020 (150)	1240 (180)	1380 (200)
Transverse [MPa (ksi)]	40 (5.8)	41 (6)	30 (4.3)
Ultimate tensile strain			
Longitudinal	2.3	0.9	1.8
Transverse	0.4	0.4	0.5

Other Fiber Reinforcement Materials

Glass, carbon, and the aramids are the most common fiber reinforcements incorporated in polymer matrices. Other fiber materials that are used to much lesser degrees are boron, silicon carbide, and aluminum oxide; tensile moduli, tensile strengths, specific strengths, and specific moduli of these materials in fiber form are contained in table. Boron fiber-reinforced polymer composites have been used in military aircraft components, helicopter rotor blades, and some sporting goods. Silicon carbide and aluminum oxide fibers are utilized in tennis rackets, circuit boards, military armor, and rocket nose cones.

Polymer Matrix Materials

In addition, the matrix often determines the maximum service temperature, since it normally softens, melts, or degrades at a much lower temperature than the fiber reinforcement.

The most widely utilized and least expensive polymer resins are the polyesters and vinyl esters;[2] these matrix materials are used primarily for glass fiber-reinforced composites. A large number of resin formulations provide a wide range of properties for these polymers. The epoxies are more expensive and, in addition to commercial applications, are also utilized extensively in PMCs for aerospace applications; they have better mechanical properties and resistance to moisture than the polyesters and vinyl resins. For high-temperature applications, polyimide resins are employed; their continuous-use, upper-temperature limit is approximately 230 °C (450 °F). Finally, high-temperature thermoplastic resins offer the potential to be used in future aerospace applications; such materials include polyetheretherketone (PEEK), poly(phenylene sulfide) (PPS), and polyetherimide (PEI).

Metal-matrix Composites

As the name implies, for metal-matrix composites (MMCs) the matrix is a ductile metal. These materials may be utilized at higher service temperatures than their base metal counterparts; furthermore, the reinforcement may improve specific stiffness, specific strength, abrasion resistance, creep resistance, thermal conductivity, and dimensional stability. Some of the advantages of these materials over the polymer-matrix composites include higher operating temperatures, nonflammability, and greater resistance to degradation by organic fluids. Metal-matrix composites are much more expensive than PMCs, and, therefore, their (MMC) use is somewhat restricted.

The superalloys, as well as alloys of aluminum, magnesium, titanium, and copper, are employed as matrix materials. The reinforcement may be in the form of particulates, both continuous and discontinuous fibers, and whiskers; concentrations normally range between 10 and 60 vol%. Continuous fiber materials include carbon, silicon carbide, boron, aluminum oxide, and the refractory metals. On the other hand, discontinuous reinforcements consist primarily of silicon carbide whiskers, chopped fibers of aluminum oxide and carbon, and particulates of silicon carbide and aluminum oxide. In a sense, the cermets fall within this MMC scheme. Table presented the properties of several common metal-matrix, continuous and aligned fiber-reinforced composites.

Some matrix–reinforcement combinations are highly reactive at elevated temperatures. Consequently, composite degradation may be caused by high-temperature processing or by subjecting the MMC to elevated temperatures during service. This problem is commonly resolved either by applying a protective surface coating to the reinforcement or by modifying the matrix alloy composition.

Normally the processing of MMCs involves at least two steps: consolidation or synthesis (i.e., introduction of reinforcement into the matrix), followed by a shaping operation. A host of consolidation techniques are available, some of which are relatively sophisticated; discontinuous fiber MMCs are amenable to shaping by standard metal-forming operations (e.g., forging, extrusion, rolling).

Automobile manufacturers have recently begun to use MMCs in their products. For example, some engine components have been introduced consisting of an aluminum-alloy matrix that is reinforced with aluminum oxide and carbon fibers; this MMC is light in weight and resists wear and thermal distortion. Metal-matrix composites are also employed in driveshafts (that have higher rotational speeds and reduced vibrational noise levels), extruded stabilizer bars, and forged suspension and transmission components.

The aerospace industry also uses MMCs. Structural applications include advanced aluminum alloy metal-matrix composites; boron fibers are used as the reinforcement for the Space Shuttle Orbiter, and continuous graphite fibers for the Hubble Telescope.

Table: Properties of several metal-matrix composites reinforced with continuous and aligned fibers.

Fiber	Matrix	Fiber Content (vol%)	Density (g/cm³)	Longitudinal Tensile Modulus (GPa)	Longitudinal Tensile Strength (MPa)
Carbon	6061 Al	41	2.44	320	620
Boron	6061 Al	48	—	207	1515
SiC	6061 Al	50	2.93	230	1480
Alumina	380.0 Al	24	—	120	340
Carbon	AZ31 Mg	38	1.83	300	510
Borsic	Ti	45	3.68	220	1270

The high-temperature creep and rupture properties of some of the superalloys (Ni-and Co-based alloys) may be enhanced by fiber reinforcement using refractory metals such as tungsten. Excellent high-temperature oxidation resistance and impact strength are also maintained. Designs incorporating these composites permit higher operating temperatures and better efficiencies for turbine engines.

Ceramic-matrix Composites

Ceramic materials are inherently resilient to oxidation and deterioration at elevated temperatures; were it not for their disposition to brittle fracture, some of these materials would be ideal candidates for use in high-temperature and severe-stress applications, specifically for components in automobile and aircraft gas turbine engines. Fracture toughness values for ceramic materials are low and typically lie between 1 and 5 MPa $\sqrt{m}$ (0.9 and 4.5 ksi $\sqrt{in}$). By way of contrast, K_{lc} values for most metals are much higher (15 to greater than 150 MPa $\sqrt{m}$ [14 to > 140 ksi $\sqrt{in}$]).

The fracture toughnesses of ceramics have been improved significantly by the development of a new generation of ceramic-matrix composites (CMCs)— particulates, fibers, or whiskers of one ceramic material that have been embedded into a matrix of another ceramic. Ceramic-matrix composite materials have extended fracture toughnesses to between about 6 and 20 MPa $\sqrt{m}$ (5.5 and 18 ksi $\sqrt{in}$).

In essence, this improvement in the fracture properties results from interactions between advancing cracks and dispersed phase particles. Crack initiation normally occurs with the matrix phase, whereas crack propagation is impeded or hindered by the particles, fibers, or whiskers. Several techniques are utilized to retard crack propagation.

One particularly interesting and promising toughening technique employs a phase transformation to arrest the propagation of cracks and is aptly termed transformation toughening. Small particles of partially stabilized zirconia are dispersed within the matrix material, often Al_2O_3 or ZrO_2

itself. Typically, CaO, MgO, Y_2O_3 and CeO are used as stabilizers. Partial stabilization allows retention of the metastable tetragonal phase at ambient conditions rather than the stable monoclinic phase; these two phases are noted on the ZrO_2–$ZrCaO_3$ phase diagram. The stress field in front of a propagating crack causes these metastably retained tetragonal particles to undergo transformation to the stable monoclinic phase. Accompanying this transformation is a slight particle volume increase, and the net result is that compressive stresses are established on the crack surfaces near the crack tip that tend to pinch the crack shut, thereby arresting its growth. This process is demonstrated schematically in figure.

Other recently developed toughening techniques involve the utilization of ceramic whiskers, often SiC or Si_3N_4. These whiskers may inhibit crack propagation by (1) deflecting crack tips, (2) forming bridges across crack faces, (3) absorbing energy during pull-out as the whiskers debond from the matrix, and (4) causing a redistribution of stresses in regions adjacent to the crack tips.

In general, increasing fiber content improves strength and fracture toughness; this is demonstrated in table for SiC whisker-reinforced alumina. Furthermore, there is a considerable reduction in the scatter of fracture strengths for whisker-reinforced ceramics relative to their unreinforced counterparts. In addition, these CMCs exhibit improved high-temperature creep behavior and resistance to thermal shock (i.e., failure resulting from sudden changes in temperature).

Ceramic-matrix composites may be fabricated using hot pressing, hot isostatic pressing, and liquid phase sintering techniques. Relative to applications, SiC whisker-reinforced aluminas are being utilized as cutting tool inserts for machining hard metal alloys; tool lives for these materials are greater than for cemented carbides.

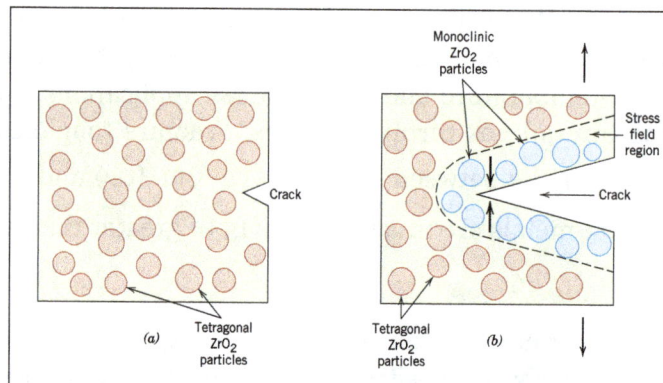

Schematic demonstration of transformation toughening. (a) A crack prior to inducement of the ZrO_2 particle phase transformation. (b) Crack arrestment due to the stressinduced phase transformation.

Carbon–Carbon Composites

One of the most advanced and promising engineering material is the carbon fiberreinforced carbon-matrix composite, often termed a carbon–carbon composite; as the name implies, both reinforcement and matrix are carbon. These materials are relatively new and expensive and, therefore, are not currently being utilized extensively. Their desirable properties include high-tensile moduli and tensile strengths that are retained to temperatures in excess of 2000 °C (3630 °F), resistance to creep, and relatively large fracture toughness values. Furthermore, carbon–carbon composites have low coefficients of thermal expansion and relatively high thermal conductivities; these characteristics, coupled with high strengths, give rise to a relatively low susceptibility to thermal shock. Their major drawback is a propensity to high temperature oxidation.

The carbon–carbon composites are employed in rocket motors, as friction materials in aircraft and high-performance automobiles, for hot-pressing molds, in components for advanced turbine engines, and as ablative shields for re-entry vehicles.

Table: Room temperature fracture strengths and fracture toughnesses for various SiC whisker contents in Al_2O_3.

Whisker Content (vol%)	Fracture Strength (MPa)	Fracture Toughness (MPa)
0	—	4.5
10	455 ± 55	7.1
20	655 ± 135	7.5–9.0
40	850 ± 130	6.0

Processing of Fiber-reinforced Composites

The primary reason that these composite materials are so expensive is the relatively complex processing techniques that are employed. Preliminary procedures are similar to those used for carbon-fiber, polymer-matrix composites. That is, the continuous carbon fibers are laid down having the desired two-or three dimensional pattern; these fibers are then impregnated with a liquid polymer resin, often a phenolic; the work piece is next formed into the final shape, and the resin is allowed to cure. At this time the matrix resin is "pyrolyzed," that is, converted into carbon by heating in an inert atmosphere; during pyrolysis, molecular components consisting of oxygen, hydrogen, and nitrogen are driven off, leaving behind large carbon chain molecules. Subsequent heat treatments at higher temperatures will cause this carbon matrix to densify and increase in strength. The resulting composite, then, consists of the original carbon fibers that remained essentially unaltered, which are contained in this pyrolyzed carbon matrix.

Hybrid Composites

A relatively new fiber-reinforced composite is the hybrid, which is obtained by using two or more different kinds of fibers in a single matrix; hybrids have a better allaround combination of properties than composites containing only a single fiber type. A variety of fiber combinations and matrix materials are used, but in the most common system, both carbon and glass fibers are incorporated into a polymeric resin. The carbon fibers are strong and relatively stiff and provide a low-density reinforcement; however, they are expensive. Glass fibers are inexpensive and lack the stiffness of carbon. The glass–carbon hybrid is stronger and tougher, has a higher impact resistance, and may be produced at a lower cost than either of the comparable all-carbon or all-glass reinforced plastics.

There are a number of ways in which the two different fibers may be combined, which will ultimately affect the overall properties. For example, the fibers may all be aligned and intimately mixed with one another; or laminations may be constructed consisting of layers, each of which consists of a single fiber type, alternating one with another. In virtually all hybrids the properties are anisotropic.

When hybrid composites are stressed in tension, failure is usually noncatastrophic (i.e., does not occur suddenly). The carbon fibers are the first to fail, at which time the load is transferred to the glass fibers. Upon failure of the glass fibers, the matrix phase must sustain the applied load. Eventual composite failure concurs with that of the matrix phase.

Principal applications for hybrid composites are lightweight land, water, and air transport structural components, sporting goods, and lightweight orthopedic components.

Processing of Fiber-reinforced Composites

To fabricate continuous fiber-reinforced plastics that meet design specifications, the fibers should be uniformly distributed within the plastic matrix and, in most instances, all oriented in virtually the same direction.

Pultrusion

Pultrusion is used for the manufacture of components having continuous lengths and a constant cross-sectional shape (i.e., rods, tubes, beams, etc.). With this technique, illustrated schematically in figure, continuous fiber rovings, or tows, are first impregnated with a thermosetting resin; these are then pulled through a steel die that preforms to the desired shape and also establishes the resin/fiber ratio. The stock then passes through a curing die that is precision machined so as to impart the final shape; this die is also heated to initiate curing of the resin matrix. A pulling device draws the stock through the dies and also determines the production speed. Tubes and hollow sections are made possible by using center mandrels or inserted hollow cores. Principal reinforcements are glass, carbon, and aramid fibers, normally added in concentrations between 40 and 70 vol%. Commonly used matrix materials include polyesters, vinyl esters, and epoxy resins.

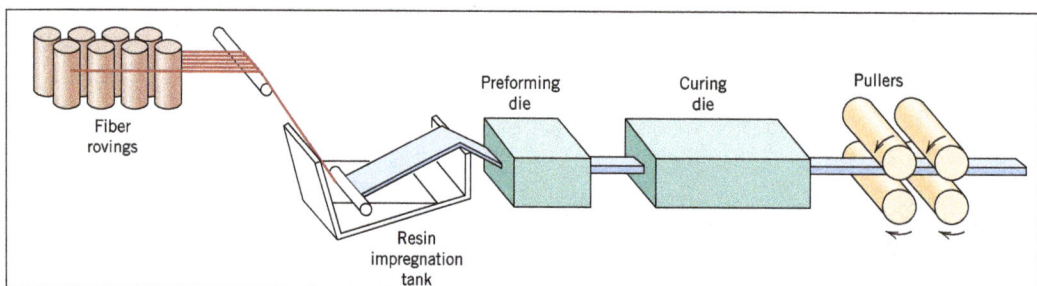

Schematic diagram showing the pultrusion process.

Pultrusion is a continuous process that is easily automated; production rates are relatively high, making it very cost effective. Furthermore, a wide variety of shapes are possible, and there is really no practical limit to the length of stock that may be manufactured.

Prepreg Production Processes

Prepreg is the composite industry's term for continuous fiber reinforcement preimpregnated with a polymer resin that is only partially cured. This material is delivered in tape form to the manufacturer, who then directly molds and fully cures the product without having to add any resin. It is probably the composite material form most widely used for structural applications.

The prepregging process, represented schematically for thermoset polymers in figure, begins by collimating a series of spool-wound continuous fiber tows. These tows are then sandwiched and pressed between sheets of release and carrier paper using heated rollers, a process termed "calendering." The release paper sheet has been coated with a thin film of heated resin solution of relatively low viscosity so as to provide for its thorough impregnation of the fibers. A "doctor blade" spreads the resin into a film of uniform thickness and width. The final prepreg product—the thin tape consisting of continuous and aligned fibers embedded in a partially cured resin—is prepared for packaging by winding onto a cardboard core. As shown in figure, the release paper sheet is removed as the impregnated tape is spooled. Typical tape thicknesses range between 0.08 and 0.25 mm $\left(3\times10^{-3} \text{ and } 10^{-2} \text{ in}\right)$ tape widths range between 25 and 1525 mm (1 and 60 in.), whereas resin content usually lies between about 35 and 45 vol%.

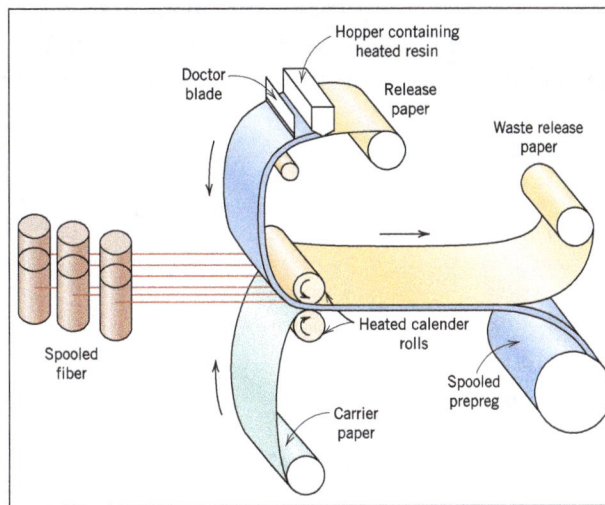

Schematic diagram illustrating the production of prepreg tape using a thermoset polymer.

At room temperature the thermoset matrix undergoes curing reactions; therefore, the prepreg is stored at 0 °C (32 °F) or lower. Also, the time in use at room temperature (or "out-time") must be minimized. If properly handled, thermoset prepregs have a lifetime of at least six months and usually longer.

Both thermoplastic and thermosetting resins are utilized; carbon, glass, and aramid fibers are the common reinforcements.

Actual fabrication begins with the "lay-up"—laying of the prepreg tape onto a tooled surface. Normally a number of plies are laid up (after removal from the carrier backing paper) to provide the desired thickness. The lay-up arrangement may be unidirectional, but more often the fiber orientation is alternated to produce a cross-ply or angle-ply laminate. Final curing is accomplished by the simultaneous application of heat and pressure.

The lay-up procedure may be carried out entirely by hand (hand lay-up), wherein the operator both cuts the lengths of tape and then positions them in the desired orientation on the tooled surface. Alternately, tape patterns may be machine cut, than hand laid. Fabrication costs can be further reduced by automation of prepreg lay-up and other manufacturing procedures, which virtually eliminates the need for hand labor. These automated methods are essential for many applications of composite materials to be cost effective.

Filament Winding

Filament winding is a process by which continuous reinforcing fibers are accurately positioned in a predetermined pattern to form a hollow (usually cylindrical) shape. The fibers either as individual strands or as tows, are first fed through a resin bath and then are continuously wound onto a mandrel, usually using automated winding equipment. After the appropriate number of layers have been applied, curing is carried out either in an oven or at room temperature, after which the mandrel is removed. As an alternative, narrow and thin prepregs (i.e., tow pregs) 10 mm or less in width may be filament wound.

Various winding patterns are possible (i.e., circumferential, helical, and polar) to give the desired mechanical characteristics. Filament-wound parts have very high strength-to-weight ratios. Also, a high degree of control over winding uniformity and orientation is afforded with this technique. Furthermore, when automated, the process is most economically attractive. Common filament-wound structures include rocket motor casings, storage tanks and pipes, and pressure vessels.

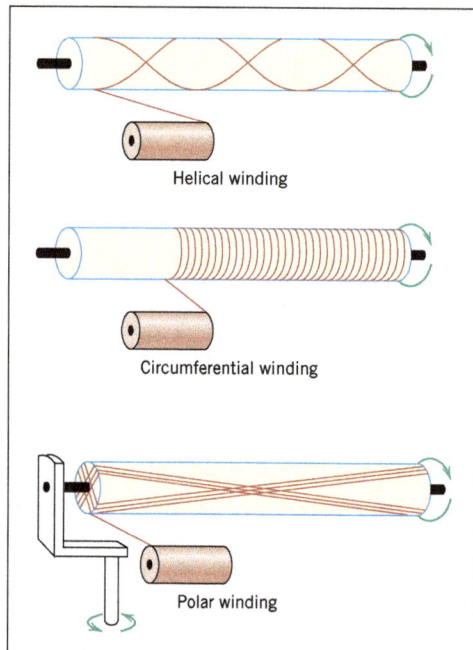

Schematic representations of helical, circumferential, and polar filament winding techniques.

Manufacturing techniques are now being used to produce a wide variety of structural shapes that are not necessarily limited to surfaces of revolution (e.g., I-beams). This technology is advancing very rapidly because it is very cost effective.

Structural Composites

A structural composite is normally composed of both homogeneous and composite materials, the properties of which depend not only on the properties of the constituent materials but also on the geometrical design of the various structural elements. Laminar composites and sandwich panels are two of the most common structural composites; only a relatively superficial examination is offered here for them.

Laminar Composites

A laminar composite is composed of two-dimensional sheets or panels that have a preferred high-strength direction such as found in wood and continuous and aligned fiber-reinforced plastics. The layers are stacked and subsequently cemented together such that the orientation of the high-strength direction varies with each successive layer. For example, adjacent wood sheets in plywood are aligned with the grain direction at right angles to each other. Laminations may also be constructed using fabric material such as cotton, paper, or woven glass fibers embedded in a plastic matrix. Thus a laminar composite has relatively high strength in a number of directions in the two-dimensional plane; however, the strength in any given direction is, of course, lower than it would be if all the fibers were oriented in that direction. One example of a relatively complex laminated structure is the modern ski.

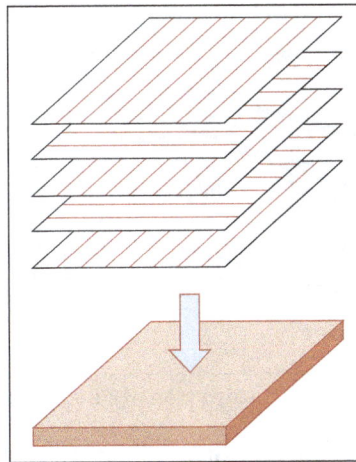

The stacking of successive oriented, fiber-reinforced layers for a laminar composite.

Sandwich Panels

Sandwich panels, considered to be a class of structural composites, are designed to be light-weight beams or panels having relatively high stiffnesses and strengths. A sandwich panel consists of two outer sheets, or faces, that are separated by and adhesively bonded to a thicker core. The outer sheets are made of a relatively stiff and strong material, typically aluminum alloys, fiber-reinforced plastics, titanium, steel, or plywood; they impart high stiffness and strength to the structure, and must be thick enough to withstand tensile and compressive stresses that result from loading. The core material is lightweight, and normally has a low modulus of elasticity. Core materials typically fall within three categories: rigid polymeric foams (i.e., phenolics, epoxy, polyurethanes), wood (i.e., balsa wood), and honeycombs.

Structurally, the core serves several functions. First of all, it provides continuous support for the faces. In addition, it must have sufficient shear strength to withstand transverse shear stresses, and also be thick enough to provide high shear stiffness (to resist buckling of the panel). (It should be noted that tensile and compressive stresses on the core are much lower than on the faces.)

Another popular core consists of a "honeycomb" structure—thin foils that have been formed into interlocking hexagonal cells, with axes oriented perpendicular to the face planes; figure shows a cutaway view of a honeycomb core sandwich panel. The honeycomb material is normally either an aluminum alloy or aramid polymer. Strength and stiffness of honeycomb structures depend on cell size, cell wall thickness, and the material from which the honeycomb is made.

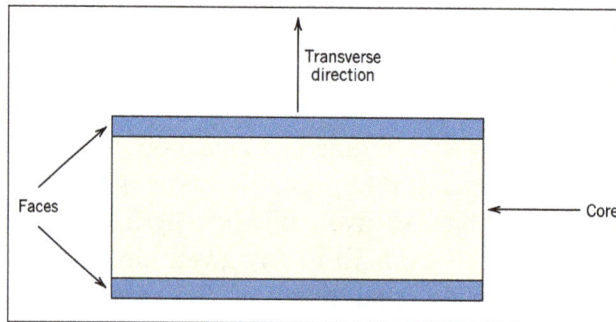

Schematic diagram showing the cross section of a sandwich panel.

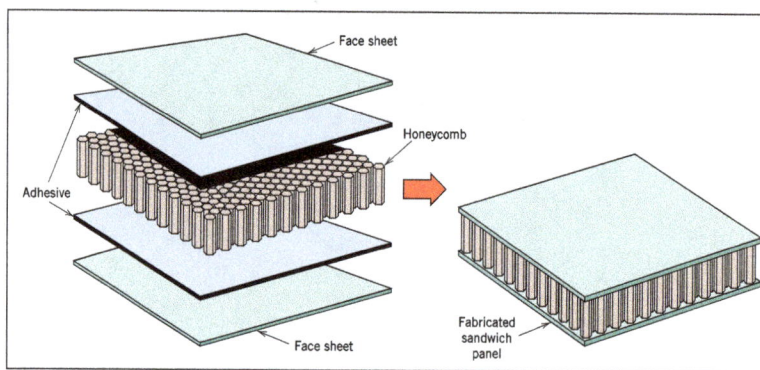

Schematic diagram showing the construction of a honeycomb core sandwich panel.

Sandwich panels are used in a wide variety of applications including roofs, floors, and walls of buildings; and, in aerospace and aircraft (i.e., for wings, fuselage, and tailplane skins).

Advantages of Composites

1. Light Weight - Composites are light in weight, compared to most woods and metals. Their lightness is important in automobiles and aircraft, for example, where less weight means better fuel efficiency (more miles to the gallon). People who design airplanes are greatly concerned with weight, since reducing a craft's weight reduces the amount of fuel it needs and increases the speeds it can reach. Some modern airplanes are built with more composites than metal including the new Boeing 787, Dreamliner.

2. High Strength - Composites can be designed to be far stronger than aluminum or steel. Metals are equally strong in all directions. But composites can be engineered and designed to be strong in a specific direction.

3. Strength Related to Weight - Strength-to-weight ratio is a material's strength in relation to how much it weighs. Some materials are very strong and heavy, such as steel. Other materials can be strong and light, such as bamboo poles. Composite materials can be designed to be both strong and light. This property is why composites are used to build airplanes—which need a very high strength material at the lowest possible weight. A composite can be made to resist bending in one direction, for example. When something is built with metal, and greater strength is needed in one direction, the material usually must be made thicker,

which adds weight. Composites can be strong without being heavy. Composites have the highest strength-to-weight ratios in structures today.

4. Corrosion Resistance - Composites resist damage from the weather and from harsh chemicals that can eat away at other materials. Composites are good choices where chemicals are handled or stored. Outdoors, they stand up to severe weather and wide changes in temperature.

5. High-Impact Strength - Composites can be made to absorb impacts—the sudden force of a bullet, for instance, or the blast from an explosion. Because of this property, composites are used in bulletproof vests and panels, and to shield airplanes, buildings, and military vehicles from explosions.

6. Design Flexibility - Composites can be molded into complicated shapes more easily than most other materials. This gives designers the freedom to create almost any shape or form. Most recreational boats today, for example, are built from fiberglass composites because these materials can easily be molded into complex shapes, which improve boat design while lowering costs. The surface of composites can also be molded to mimic any surface finish or texture, from smooth to pebbly.

7. Part Consolidation - A single piece made of composite materials can replace an entire assembly of metal parts. Reducing the number of parts in a machine or a structure saves time and cuts down on the maintenance needed over the life of the item.

8. Dimensional Stability - Composites retain their shape and size when they are hot or cool, wet or dry. Wood, on the other hand, swells and shrinks as the humidity changes. Composites can be a better choice in situations demanding tight fits that do not vary. They are used in aircraft wings, for example, so that the wing shape and size do not change as the plane gains or loses altitude.

9. Nonconductive - Composites are nonconductive, meaning they do not conduct electricity. This property makes them suitable for such items as electrical utility poles and the circuit boards in electronics. If electrical conductivity is needed, it is possible to make some composites conductive.

10. Nonmagnetic - Composites contain no metals; therefore, they are not magnetic. They can be used around sensitive electronic equipment. The lack of magnetic interference allows large magnets used in MRI (magnetic resonance imaging) equipment to perform better. Composites are used in both the equipment housing and table. In addition, the construction of the room uses composites rebar to reinforced the concrete walls and floors in the hospital.

11. Radar Transparent - Radar signals pass right through composites, a property that makes composites ideal materials for use anywhere radar equipment is operating, whether on the ground or in the air. Composites play a key role in stealth aircraft, such as the U.S. Air Force's B-2 stealth bomber, which is nearly invisible to radar.

12. Low Thermal Conductivity - Composites are good insulators—they do not easily conduct heat or cold. They are used in buildings for doors, panels, and windows where extra protection is needed from severe weather.

13. Durable - Structures made of composites have a long life and need little maintenance. We do not know how long composites last, because we have not come to the end of the life of many original composites. Many composites have been in service for half a century.

Disadvantages of Composites

1. Composite fillings wear out sooner than amalgam fillings (lasting at least five years compared with at least 10 to 15 for amalgams); in addition, they may not last as long as amalgam fillings under the pressure of chewing and particularly if used for large cavities.

2. Because of the process to apply the composite material, these fillings can take up to 20 minutes longer than amalgam fillings to place.

3. If composites are used for inlays or onlays, more than one office visit may be required.

4. Depending on location, composite materials can chip off the tooth.

5. Composite fillings can cost up to twice the cost of amalgam fillings.

Metal Matrix Composites

The composite materials which have two or more constituent parts with at least one being a metal are known as metal matrix composites. They are processed through different techniques such as solid state processing, liquid state processing and in-situ processing. The topics elaborated in this chapter will help in gaining a better perspective about these processes as well as the classification of metal matrix composites.

Metal-matrix composites are either in use or prototyping for the Space Shuttle, commercial airliners, electronic substrates, bicycles, automobiles, golf clubs, and a variety of other applications. While the vast majority are aluminum matrix composites, a growing number of applications require the matrix properties of superalloys, titanium, copper, magnesium, or iron.

Like all composites, aluminum-matrix composites are not a single material but a family of materials whose stiffness, strength, density, and thermal and electrical properties can be tailored. The matrix alloy, the reinforcement material, the volume and shape of the reinforcement, the location of the reinforcement, and the fabrication method can all be varied to achieve required properties. Regardless of the variations, however, aluminum composites offer the advantage of low cost over most other MMCs. In addition, they offer excellent thermal conductivity, high shear strength, excellent abrasion resistance, high-temperature operation, non-flammability, minimal attack by fuels and solvents, and the ability to be formed and treated on conventional equipment.

Aluminum MMCs are produced by casting, powder metallurgy, in situ development of reinforcements, and foil-and-fiber pressing techniques. Consistently high-quality products are now available in large quantities, with major producers scaling up production and reducing prices. They are applied in brake rotors, pistons, and other automotive components, as well as golf clubs, bicycles, machinery components, electronic substrates, extruded angles and channels, and a wide variety of other structural and electronic applications.

Superalloy composites reinforced with tungsten alloy fibers are being developed for components in jet turbine engines that operate temperatures above 1,830 °F.

Graphite/copper composites have tailorable properties, are useful to high temperatures in air, and provide excellent mechanical characteristics, as well as high electrical and thermal conductivity. They offer easier processing as compared with titanium, and lower density compared with steel. Ductile superconductors have been fabricated with a matrix of copper and superconducting filaments of niobium-titanium. Copper reinforced with tungsten particles or aluminum oxide particles is used in heat sinks and electronic packaging.

Titanium reinforced with silicon carbide fibers is under development as skin material for the National Aerospace Plane. Stainless steels, tool steels, and Inconel are among the matrix materials reinforced with titanium carbide particles and fabricated into draw-rings and other high-temperature, corrosion-resistant components.

Compared to monolithic metals, MMCs have:

- Higher strength-to-density ratios.

- Higher stiffness-to-density ratios.

- Better fatigue resistance.

- Better elevated temperature properties:

    ○ Higher strength.

    ○ Lower creep rate.

- Lower coefficients of thermal expansion.

- Better wear resistance.

The advantages of MMCs over polymer matrix composites are:

- Higher temperature capability.

- Fire resistance.

- Higher transverse stiffness and strength.

- No moisture absorption.

- Higher electrical and thermal conductivities.

- Better radiation resistance.

- No outgassing.

- Fabricability of whisker and particulate-reinforced MMCs with conventional metalworking equipment.

Some of the disadvantages of MMCs compared to monolithic metals and polymer matrix composites are:

- Higher cost of some material systems.

- Relatively immature technology.

- Complex fabrication methods for fiber-reinforced systems (except for casting).

- Limited service experience.

Numerous combinations of matrices and reinforcements have been tried since work on MMC began in the late 1950s. However, MMC technology is still in the early stages of development, and other important systems undoubtedly will emerge.

Reinforcements: MMC reinforcements can be divided into five major categories: continuous fibers, discontinuous fibers, whiskers, particulates, and wires. With the exception of wires, which are metals, reinforcements generally are ceramics.

Key continuous fibers include boron, graphite (carbon), alumina, and silicon carbide. Boron fibers are made by chemical vapor deposition (CVD) of this material on a tungsten core. Carbon cores have also been used. These relatively thick monofilaments are available in 4.0, 5.6, and 8.0-mil diameters. To retard reactions that can take place between boron and metals at high temperature, fiber coatings of materials such as silicon carbide or boron carbide are sometimes used.

Silicon carbide monofilaments are also made by a CVD process, using a tungsten or carbon core. A Japanese multifilament yarn, designated as silicon carbide by its manufacturer, is also commercially available. This material, however, made by pyrolysis of organometallic precursor fibers, is far from pure silicon carbide and its properties differ significantly from those of monofilament silicon carbide.

Continuous alumina fibers are available from several suppliers. Chemical compositions and properties of the various fibers are significantly different. Graphite fibers are made from two precursor materials, polyacrilonitrile (PAN) and petroleum pitch. Efforts to make graphite fibers from coal-based pitch are under way. Graphite fibers with a wide range of strengths and moduli are available.

The leading discontinuous fiber reinforcements at this time are alumina and alumina-silica. Both originally were developed as insulating materials. The major whisker material is silicon carbide. The leading U.S. commercial product is made by pyrolysis of rice hulls. Silicon carbide and boron carbide, the key particulate reinforcements, are obtained from the commercial abrasives industry. Silicon carbide particulates are also produced as a by-product of the process used to make whiskers of this material.

A number of metal wires including tungsten, beryllium, titanium, and molybdenum have been used to reinforce metal matrices. Currently, the most important wire reinforcements are tungsten wire in superalloys and superconducting materials incorporating niobium-titanium and niobium-tin in a copper matrix. The reinforcements cited above are the most important at this time. Many others have been tried over the last few decades, and still others undoubtedly will be developed in the future.

Matrix Materials and Key Composites

Numerous metals have been used as matrices. The most important have been aluminum, titanium, magnesium, and copper alloys and superalloys.

The most important MMC systems are:

- Aluminum matrix:
    - Continuous fibers: boron, silicon carbide, alumina, graphite.
    - Discontinuous fibers: alumina, alumina-silica.
    - Whiskers: silicon carbide.
    - Particulates: silicon carbide, boron carbide.
- Magnesium matrix:
    - Continuous fibers: graphite, alumina.
    - Whiskers: silicon carbide.

- ◦ Particulates: silicon carbide, boron carbide.
- Titanium matrix:
  - ◦ Continuous fibers: silicon carbide, coated boron.
  - ◦ Particulates: titanium carbide.
- Copper matrix:
  - ◦ Continuous fibers: graphite, silicon carbide.
  - ◦ Wires: niobium-titanium, niobium-tin.
  - ◦ Particulates: silicon carbide, boron carbide, titanium carbide.
- Superalloy matrices:
  - ◦ Wires: tungsten.

Characteristics and Design Considerations

The superior mechanical properties of MMCs drive their use. An important characteristic of MMCs, however, and one they share with other composites, is that by appropriate selection of matrix materials, reinforcements, and layer orientations, it is possible to tailor the properties of a component to meet the needs of a specific design.

For example, within broad limits, it is possible to specify strength and stiffness in one direction, coefficient of expansion in another, and so forth. This is rarely possible with monolithic materials.

Monolithic metals tend to be isotropic, that is, to have the same properties in all directions. Some processes such as rolling, however, can impart anisotropy, so that properties vary with direction. The stress-strain behavior of monolithic metals is typically elastic-plastic. Most structural metals have considerable ductility and fracture toughness.

The wide variety of MMCs have properties that differ dramatically. Factors influencing their characteristics include:

- Reinforcement properties, form, and geometric arrangement.
- Reinforcement volume fraction.
- Matrix properties, including effects of porosity.
- Reinforcement-matrix interface properties.
- Residual stresses arising from the thermal and mechanical history of the composite.
- Possible degradation of the reinforcement resulting from chemical reactions at high temperatures, and mechanical damage from processing, impact, etc.

Particulate-reinforced MMCs, like monolithic metals, tend to be isotropic. The presence of brittle reinforcements and perhaps of metal oxides, however, tends to reduce their ductility and fracture toughness. Continuing development may reduce some of these deficiencies.

The properties of materials reinforced with whiskers depend strongly on their orientation. Randomly oriented whiskers produce an isotropic material. Processes such as extrusion can orient whiskers, however, resulting in anisotropic properties. Whiskers also reduce ductility and fracture toughness.

MMCs reinforced with aligned fibers have anisotropic properties. They are stronger and stiffer in the direction of the fibers than perpendicular to them. The transverse strength and stiffness of unidirectional MMCs (materials having all fibers oriented parallel to one axis), however, are frequently great enough for use in components such as stiffeners and struts. This is one of the major advantages of MMCs over PMCs, which can rarely be used without transverse reinforcement.

Because the modulus and strength of metal matrices are significant with respect to those of most reinforcing fibers, their contribution to composite behavior is important. The stress-strain curves of MMCs often show significant nonlinearity resulting from yielding of the matrix.

Another factor that has a significant effect on the behavior of fiber-reinforced metals is the frequently large difference in coefficient of expansion between the two constituents. This can cause large residual stresses in composites when they are subjected to significant temperature changes. In fact, during cool down from processing temperatures, matrix thermal stresses are often severe enough to cause yielding. Large residual stresses can also be produced by mechanical loading.

Although fibrous MMCs may have stress-strain curves displaying some nonlinearity, they are essentially brittle materials, as are PMCs. In the absence of ductility to reduce stress concentrations, joint design becomes a critical design consideration. Numerous methods of joining MMCs have been developed, including metallurgical and polymeric bonding and mechanical fasteners.

Fabrication Methods

Fabrication methods are an important part of the design process for all structural materials, including MMCs. Considerable work is under way in this critical area. Significant improvements in existing processes and development of new ones appear likely.

Current methods can be divided into two major categories, primary and secondary. Primary fabrication methods are used to create the MMC from its constituents. The resulting material may be in a form that is close to the desired final configuration, or it may require considerable additional processing, called secondary fabrication, such as forming, rolling, metallurgical bonding, and machining. The processes used depend on the type of reinforcement and matrix.

A critical consideration is reactions that can occur between reinforcements and matrices during primary and secondary processing at the high temperatures required to melt and form metals. These impose limitations on the kinds of constituents that can be combined by the various processes. Sometimes, barrier coatings can be successfully applied to reinforcements, allowing them to be combined with matrices that otherwise would be too reactive. For example, the application of a coating such as boron carbide permits the use of boron fibers to reinforce titanium. Potential reactions between matrices and reinforcements, even coated ones is also an important criterion in evaluating the temperatures and corresponding lengths of time to which MMCs may be subjected in service.

Relatively large-diameter monofilament fibers, such as boron and silicon carbide, have been incorporated into metal matrices by hot pressing a layer of parallel fibers between foils to create a monolayer tape. In this operation, the metal flows around the fibers and diffusion bonding occurs. The same procedure can be used to produce diffusion-bonded laminates with layers of fibers oriented in specified directions to meet stiffness and strength requirements for a particular design. In some instances, laminates are produced by hot pressing monolayer tapes in what can be considered a secondary operation.

Monolayer tapes are also produced by spraying metal plasmas on collimated fibers, followed by hot pressing. Structural shapes can be fabricated by creep and superplastic forming of laminates in a die. An alternate process is to place fibers and unbonded foils in a die and hot press the assembly.

The boron/aluminum struts used on the space shuttle are fabricated from monolayer foils wrapped around a mandrel and hot isostatically pressed to diffusion bond the foil layers together and, at the same time, to diffusion bond the composite laminate to titanium end fittings.

Composites can be made by infiltrating liquid metal into a fabric or prearranged fibrous configuration called a preform. Frequently, ceramic or organic binder materials are used to hold the fibers in position. The latter is burned off before or during infiltration. Infiltration can be carried out under vacuum, pressure, or both. Pressure infiltration, which promotes wetting of the fibers by the matrix and reduces porosity, is often called squeeze casting.

Cast MMCs now consistently offer net or net-net shape, improved stiffness and strength, and compatibility with conventional manufacturing techniques. They are also consistently lower in cost than those produced by other methods, are available from a wide range of fabricators, and offer dimensional stability in both large and small parts.

For example, Duralcan has developed its "ice cream mixer" technology and process controls to the point where it produces up to 25 million pounds per year of aluminum composite billets. Investment casting has been modified at Cercast to cast Duralcan billets into complex, net-shape parts. Pressure casting produces net shapes with exceptional properties at Alcoa, while pressureless infiltration is used at Lanxide Corp. to fabricate net-shape components.

At the current time, the most common method used to make graphite/aluminum and graphite/magnesium composites is by infiltration. Graphite yarn is first passed through a furnace to burn off any sizing that may have been applied. Next it goes through a CVD process that applies a coating of titanium and boron which promotes wetting by the matrix. Then it immediately passes through a bath or fountain of molten metal, producing an infiltrated bundle of fibers known as a "wire." Plates and other structural shapes are produced in a secondary operation by placing the wires between foils and pressing them, as is done with monofilaments. Recent development of "air stable" coatings permits use of other infiltration processes, such as casting, eliminating the need for "wires" as an intermediate step. Other approaches are under development.

A particularly important secondary fabrication method for titanium matrix composites is superplastic forming/diffusion bonding (SPF/DB). To reduce fabrication costs, continuous processes such as pultrusion and hot roll bonding are being developed.

Three basic methods are being used to make whisker and particulate-reinforced MMCs. Two use powdered metals; the other uses a liquid-metal approach, details of which are proprietary.

The two powder-metal processes differ primarily in the way the constituents are mixed. One uses a ball mill, the other employs a liquid to aid mixing, which is subsequently removed. Mixtures are then hot pressed into billets.

Secondary processes are similar to those for monolithic metals, including rolling, extrusion, spinning, forging, creep-forming, and machining. The latter poses some difficulties because the reinforcements are very hard.

Classification of Metal Matrix Composites

MMCs are classified into different categories depending upon their matrix materials. Some examples of most commonly used metallic matrix configurations are:

- Aluminum-based composites; aluminum as matrix can be either cast alloy or wrought alloy (i.e., AlMgSi, AlMg, AlCuSiMn, AlZnMgCu, AlCu, AlSiCuMg).

- Magnesium-based composites.

- Titanium-based composites.

- Copper-based composites.

- Super alloy-based composites.

In figure shows the usage volume of different matrix materials in MMCs. As seen, aluminum is the most commonly sued matrix material in MMCs.

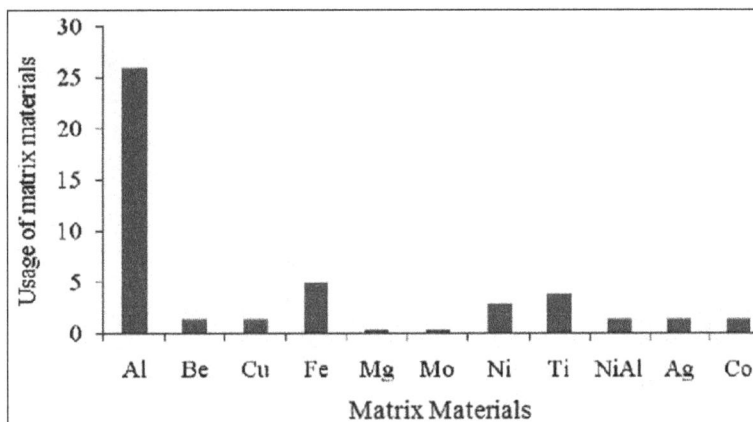

Usage of matrix materials in MMCs.

Aluminum–matrix composites are most commonly studied MMC as they are widely used in the automotive and aerospace industries. Reinforcement compounds such as SiC, Al_2O_3, and B_4C can be mixed easily and effectively in molten aluminum. Magnesium–matrix composites have similar advantages, but due to limitations in fabrication and lower thermal conductivity, they are not widely used as compared with aluminum-based MMCs. Magnesium–matrix composites have been developed for the space industry thank to the low density of magnesium and its alloys. Titanium alloys are used as matrix material in fabricating MMCs due to their good strength at elevated temperatures and excellent corrosion resistance. Compared with aluminum, titanium alloys hold their strength at higher temperature, which is advantageous in manufacturing aircraft and missile structures, whose operating speeds are very high. However, their main problem lies with processing of highly reactive titanium with reinforcement materials. Fiber-based titanium composites are widely used in developing aircraft structures. In terms of thermal conductivity and high- temperature strength properties, copper–matrix composites are superior compared with other MMCs. Super alloys are commonly used as candidate materials for manufacturing gas turbine blades, where they operate at higher temperatures and speeds.

Aluminum-based MMCs

Aluminum-based MMCs are the most commonly used MMC in the automotive and aerospace applications. This is mainly due to its unique properties like greater strength, improved stiffness, reduced density, improved temperature properties, controlled thermal expansion and improved wear resistance. In aluminum-based MMCs, one of the constituent is an aluminum/aluminum alloy (i.e., Al–Si, Al–Cu, Al–Si–Mg alloys), which forms percolating network and is termed as matrix phase. The other constituent is embedded in this aluminum/aluminum alloy matrix and serves as reinforcement, which is usually non-metallic and common ceramic such as SiC, Al_2O_3, C, B, B_4C, AlN, and BN. Figure shows the microstructure of a SiC-particle reinforced aluminum–matrix composite material. Properties of aluminum-based MMCs can be tailored by varying the nature of constituents and their volume fraction.

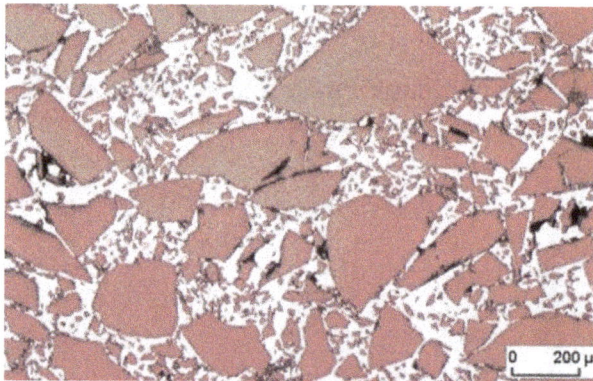

SiC-particle reinforced aluminum-matrix composite material (SiC-volume content: 70%).

Aluminum-based MMCs are intended to substitute monolithic materials including aluminum alloys, ferrous alloys, titanium alloys and polymer based composites in several applications. The major advantages of aluminum-based MMCs compared to unreinforced materials are as follows:

- Greater strength.

- Improved stiffness.

- Reduced density (weight).

- Improved high-temperature properties.

- Controlled thermal expansion coefficient.

- Thermal/heat management.

- Enhanced and tailored electrical performance.

- Improved abrasion and wear resistance.

- Control of mass (especially in reciprocating applications).

- Improved damping capabilities.

These advantages can be quantified for better appreciation. For example, elastic modulus of pure aluminum can be enhanced from 70 GPa to 240 GPa by reinforcing with 60 vol% continuous aluminum fiber. On the other hand, incorporation of 60 vol% alumina fiber in pure aluminum leads to decrease in the coefficient of thermal expansion from 24 ppm/°C to 7 ppm/°C. Similarly it is possible to process Al–9% Si–20 vol% SiCp composites having wear resistance equivalent or better than that of gray cast iron. All these examples illustrate that it is possible to alter several technological properties of aluminum/aluminum alloy by more than two to three orders of magnitude by incorporating appropriate reinforcement in suitable volume fraction.

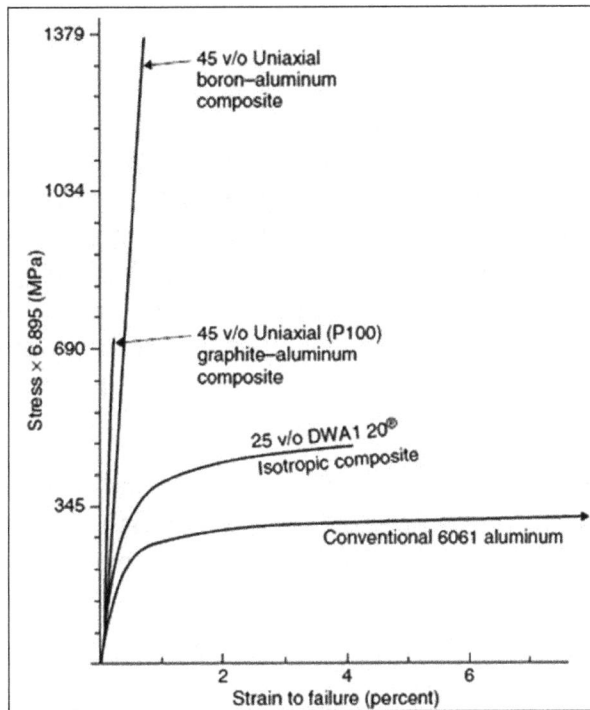

Comparative stress–strain curves of a 6061 aluminum alloy and the aluminum-based composites.

In figure superimposes curves for the three primary aluminum-based MMCs, ceramic particulate, boron filament, and graphite tow reinforcements, on the curve for conventional 6061-T_6. Although these curves are somewhat idealized, the two uniaxially reinforced continuous fiber composites are markedly stiffer and stronger but also show very low ductility-to-fracture. The DWA1 20, ceramic particulate-reinforced composite (25 vol% 6061-T_6) shows a more conventional stress–strain response (exhibiting ductility), but shows mechanical properties intermediate between those of the unreinforced material and those of the continuous fiber composites. The exact positions and shapes of the composite stress–strain curves can be varied by matrix heat treatment, reinforcement volume percent (vol%), or reinforcement orientation effects. It must also be kept in mind that all continuously reinforced composite material exhibits anisotropic characteristics. In figure, if the transverse direction stress–strain curves for the boron–aluminum material were to be included, it would be relegated to the extreme lower left-hand corner of the chart: strength 152 MPa, modulus ~145 GPa, $\mathring{a}_r \sim 0.4\%$. Similarly, the graphite–aluminum composite transverse stress–strain response shows even lower mechanical characteristics.

Over the years, aluminum–matrix composites have been used in numerous structural, non-structural and functional applications in different engineering sectors. Driving force for the utilization of aluminum–matrix composites in these sectors include performance, economic and environmental benefits. The key benefits of aluminum-based MMCs composites in transportation sector are lower fuel consumption, less noise and lower airborne emissions. With increasing strict environmental regulations and emphasis on improved fuel economy, use of aluminum-based MMCs in transportation sector will be inevitable and desirable in the coming years. The aluminum-based MMCs are very attractive for their isotropic mechanical properties (higher than their unreinforced alloys) and their low costs (cheap processing routes and low prices of some of the discontinuous reinforcement such as SiC particles or Al_2O_3 short fibers).

Magnesium-based MMCs

The increasing demand for lightweight and high-performance materials is likely to increase the need for magnesium-based MMCs. The MMCs based on magnesium alloys, in particular Mg–Al system, are excellent candidates for engineering light structure materials, and have great potential in civic, military and aerospace applications. The potential applications of magnesium–matrix composites in the automotive industry include their use in disk rotors, piston ring grooves, gears, gearbox bearings, connecting rods, and shift forks. One of the drawbacks of magnesium MMC is higher production costs due to more complicated manufacturing processes. To address this, the use of low cost materials, alloys and reinforcements can provide room for this class of low density materials into the market.

The development of a wide range of reinforcing materials and new processing techniques are among top interests in high-performance magnesium materials. The magnesium-based MMCs uni-directionally reinforced with continuous carbon fiber can readily show a bending strength of 1000 MPa with a density as low as 1.8 g cm^{-3}. The superior mechanical property can be retained at elevated temperatures of up to 350–400 °C. In some magnesium alloys, formation of a composite may be the only effective approach to strengthening. Mg–Li binary alloys with eutectic composition, for instance, are composed of HCP (α) and BCC (β) solid solution phases. The dissolution of Li into Mg causes a partial solid solution strengthening effect without the formation of any Mg–Li precipitates during the cooling process. Therefore, heat treatment is not an effective way to improve mechanical properties of these alloys. Considering this, the incorporation of thermally stable reinforcements into composite materials makes them applicable for elevated-temperature applications.

Mg–Al alloys such as AM60 and AZ91 are presently the most prevalent magnesium alloys utilized in the automotive industry. They are also the most widely studied matrix for magnesium-based composites. Other magnesium materials, such as pure magnesium, Mg–Li alloy, and Mg–Ag–Re (QE22) alloys, have also been employed as a matrix material, though less frequently. Ceramic particles are the most widely studied reinforcement for magnesium–matrix composites. Some common properties of ceramic materials make them desirable for reinforcements. These properties include low density and high levels of hardness, strength, elastic modulus, and thermal stability. However, they also have some common limitations such as low wettability, low ductility, and low compatibility with a magnesium matrix. Among the various ceramic reinforcements, SiC is the most popular one because of its relatively high wettability and its stability in magnesium melt, as compared to other ceramics.

In figures shows optical image of the as-cast AZ91D-based composite with 50 vol% $Mg_2B_2O_5$ whiskers and scanning electron micrograph of AZ91 based composite reinforced with SiC particles, respectively.

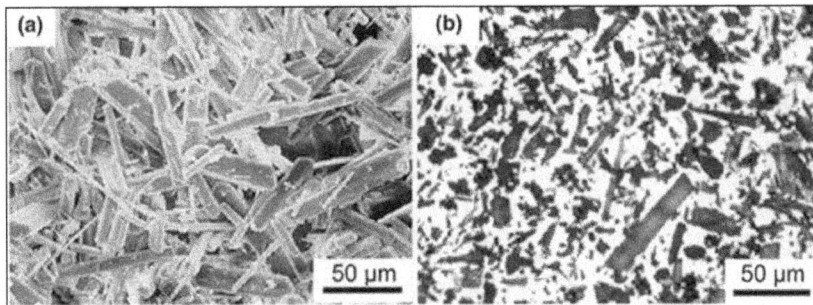

Optical image of the as-cast AZ91D-based composite with 50 vol% $Mg_2B_2O_5$ whiskers.

Copper-based MMCs

When Al_2O_3 particles are dispersed in copper matrix, unique characteristics can be achieved (i.e., high thermal and electrical conductivity, as well as high strength and excellent resistance to annealing). The applications encompass resistance welding electrodes, lead frames and electrical connectors.

The materials for electronic packaging and thermal management applications should have compatible coefficients of thermal expansion (CTE) with those of semiconductors or ceramic substrates, high thermal conduction and excellent mechanical properties. Due to the high thermal conductivity of copper and low CTE of SiC, CuSiC MMC can be made to serve as a good solution for thermal management. Semiconductors and ceramics have CTE in the range of 2–7 ppm/°C. Traditional low-CTE materials like copper/tungsten (Cu/W), copper/molybdenum (Cu/Mo), copper–Invar–copper (Cu/I/Cu) and copper–molybdenum–copper (Cu/Mo/Cu) have high densities and thermal conductivities that are little or no better than that of aluminum; Density, thermal conductivity and CTE concerns mentioned above can be copped by using copper silicon carbide (CuSiC) based MMCs. Copper silicon carbide composites provide a good compromise between thermo-mechanical properties and high conductivity. They have lower density than copper, very good thermal conductivity, low CTE and good machinability. A CuSiC based MMC heat spreader will offer high thermal conductivity between 250 W/m K and 325 W/m K and corresponding adjustable CTE between 8.0 ppm/°C and 12.5 ppm/1°C. However, the primary challenge of CuSiC manufacturing is to prevent reaction between copper and silicon carbide during high-temperature densification, which dramatically degraded the thermal conductivity.

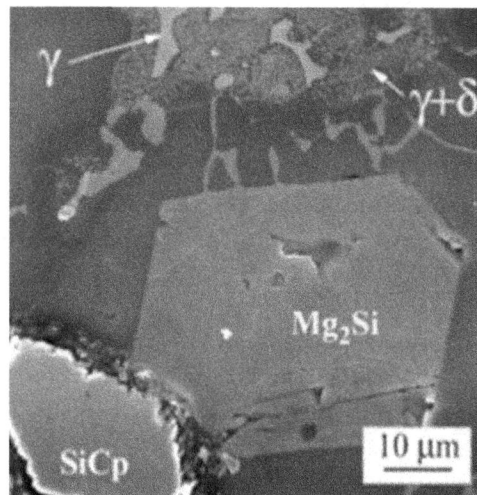

Scanning electron micrograph of AZ91 metal matrix composite reinforced with SiC particles.

Titanium-based MMCs

Ti-MMCs reinforced by continuous SiC fiber are being developed for aerospace applications in several countries, including the USA, UK, France, and China, Peng. Ti-MMCs provide outstanding mixture of stiffness, specific strength, fatigue and creep resistance at elevated temperatures. Owning to the active nature of Ti, several solid-state processing techniques have been developed to date for manufacturing Ti-MMCs, Guo and Derby, including the foil-fiber-foil (FFF) method, matrix-coated mono-tape (MCM) method and the matrix-coated fiber (MCF) method. A maximum fiber volume fraction up to 80% has been achieved with very uniform fiber distribution in MCF. The research on the MCF method has been mainly concentrated on the consolidation behavior of MCFs with the aim to optimize the processing parameters. Table compares physical and mechanical properties of a Ti alloy and a Ti-MMC. Figure shows the microstructure of an SiC reinforced titanium matrix composite.

Table: Physical and mechanical properties of a Ti alloy and a Ti-MMC.

Property	Titanium (Ti-6-2-4-2)	Ti-MMC
Density, gr/cm³	4.54	3.93
Tensile strength (longitudinal), MPa	931	1689
Young's modulus (longitudinal), GPa	117	200
Tensile strength (transverse), MPa	931	400
Young's modulus (transverse), GPa	117	145
Compression strength, MPa	931	44481
High cycle fatigue (longitudinal), R ¼ 0.1 and 30 Hz	10^7 cycles 482 MPa	10^7 cycles 531 MPa
Tensile strength at 315.5 °C, MPa	552	1379

Super Alloy-based Mmcs

The primary application for super alloy matrix composites is gas turbine blades. By enhancing the material operating temperatures and stresses of turbine blades, an increase in the performance and a reduction in the operating cost can be achieved. Several composites have been chosen for developments of the next generation of turbine blade materials. Refractory alloy wire reinforced super alloys have been investigated, for instance, for elevated-temperature applications. Solid state diffusion bonding and liquid phase infiltration techniques have been used to produce composite samples. Mono-layers tape has been produced using techniques and equipment similar to that used for Ti–matrix composites.

Microstructures of a SiC reinforced titanium matrix composite.

In the field of super alloy matrix composites, there have been several recent extensive reviews of progress in the development of Nb-silicide-based composites shows the microstructure of a Nb-silicide-based composites). In the elevated-temperature applications, for the next generation of jet engines, Nb- and Mo-silicide-based composites are among the best candidates. The melting points of the silicide containing composites based on these systems are in excess of 1750 °C. Densities of the Nb-silicide-based composites are in the range of 6.6–7.2 g/cc The ambient-temperature fracture toughness of Nb-silicide-based composite systems has been reported as being above 20 MPa m$^{1/2}$ while values for the preliminary creep and oxidation properties indicate that Nb-silicide-based composites could, with further development, be integrated into blade designs with substantial payoffs in weight and cooling-air savings, relative to fourth generation super alloy blade designs.

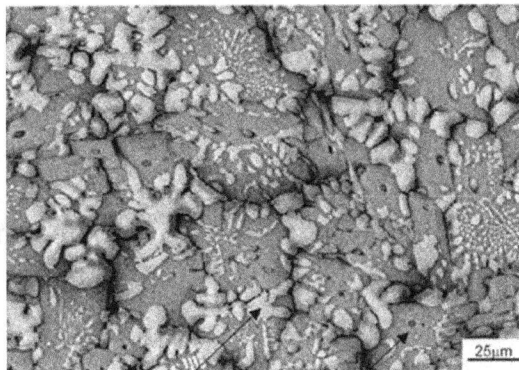

Backscattered electron image of a Nb-silicide composite (a two-phase composite of Nb and Nb$_3$Si).

Oxide (ceramic)-fiber-reinforced super alloys are another important class of composites being studied for high-temperature applications. Single crystal aluminum-oxide super alloys based composites are produced commercially for research activity purposes. Productions rates are relatively slow and fiber unit cost is quite high and composite properties suitable for industrial applications have yet to be demonstrated.

Processing of Metal Matrix Composites

Processing of metal matrix composites (MMC) can be classified into three main categories:

- Solid State Processing
- Liquid State Processing
- In-Situ Processing

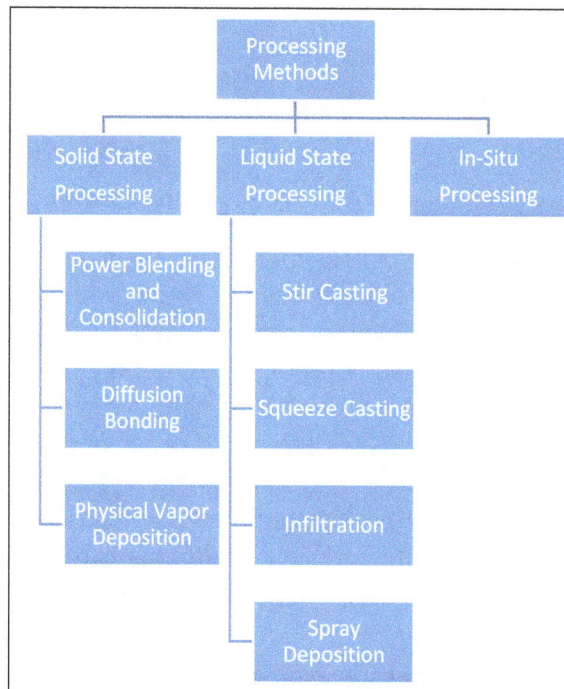

Solid State Processing

Solid state fabrication of Metal Matrix Composites is the process, in which Metal Matrix Composites are formed as a result of bonding matrix metal and dispersed phase due to mutual diffusion occurring between them in solid states at elevated temperature and under pressure.

Low temperature of solid state fabrication process (as compared to Liquid state fabrication of Metal Matrix Composites) depresses undesirable reactions on the boundary between the matrix and dispersed (reinforcing) phases.

Metal Matrix Composites may be deformed also after sintering operation by rolling, Forging, pressing, Drawing or Extrusion. The deformation operation may be either cold (below the recrystallization temperature) or hot (above the recrystallyzation temperature).

Deformation of sintered composite materials with dispersed phase in form of short fibers results in a preferred orientation of the fibers and anisotropy of the material properties (enhanced strength along the fibers orientation).

There are two principal groups of solid state fabrication of Metal Matrix Composites:

- Diffusion bonding
- Sintering

Diffusion Bonding

Diffusion Bonding is a solid state fabrication method, in which a matrix in form of foils and a dispersed phase in form of long fibers are stacked in a particular order and then pressed at elevated temperature.

The finished laminate composite material has a multilayer structure.

Diffusion Bonding is used for fabrication of simple shape parts (plates, tubes).

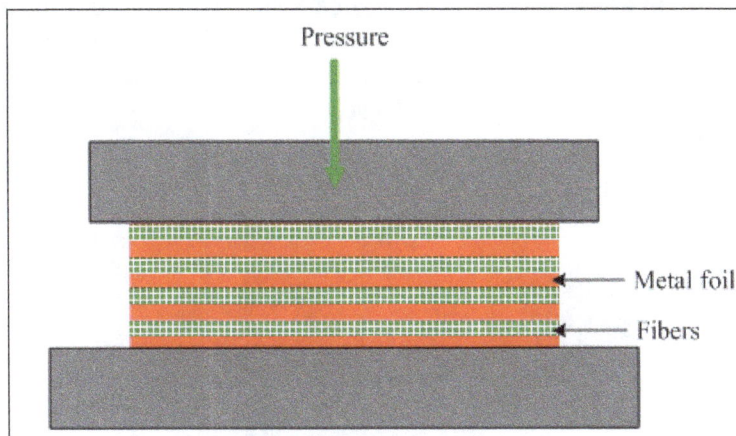

Variants of diffusion bonding are roll bonding and wire/fiber winding:

- Roll Bonding is a process of combined Rolling (hot or cold) strips of two different metals (e.g. steel and aluminum alloy) resulted in formation of a laminated composite material with a metallurgical bonding between the two layers.
- Wire/fiber Winding is a process of combined winding continuous ceramic fibers and metallic wires followed by pressing at elevated temperature.

Sintering

Sintering fabrication of Metal Matrix Composites is a process, in which a powder of a matrix metal is mixed with a powder of dispersed phase in form of particles or short fibers for subsequent compacting and sintering in solid state (sometimes with some presence of liquid).

Sintering is the method involving consolidation of powder grains by heating the "green" compact part to a high temperature below the melting point, when the material of the separate particles diffuse to the neighboring powder particles. In contrast to the liquid state fabrication of Metal Matrix Composites, sintering method allows obtaining materials containing up to 50% of dispersed phase.

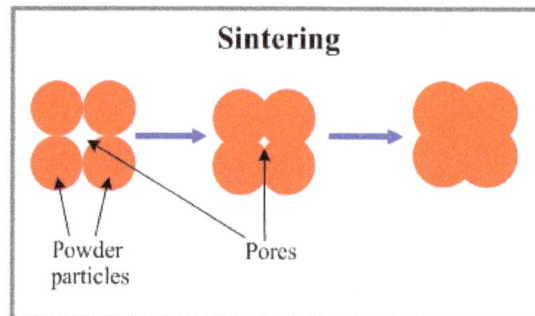

When sintering is combined with a deformation operation, the fabrication methods are called:

Hot Pressing Fabrication of Metal Matrix Composites

Hot Pressing Fabrication of Metal Matrix Composites – sintering under a unidirectional pressure applied by a hot press.

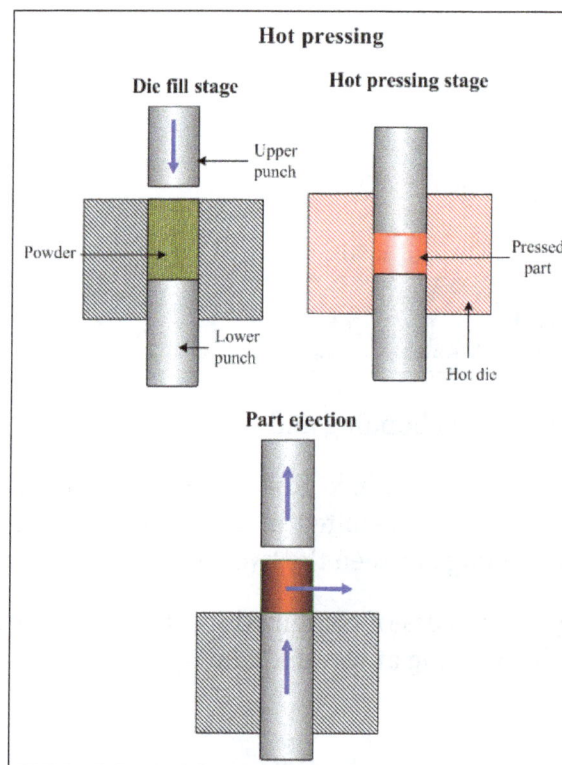

Hot Isostatic Pressing Fabrication of Metal Matrix Composites

Hot Isostatic Pressing Fabrication of Metal Matrix Composites – sintering under a pressure applied from multiple directions through a liquid or gaseous medium surrounding the compacted part and at elevated temperature.

Hot isostatic pressing

Pressure

Pressurized gas (argon)

Pressed part

Steel can container

Heated chamber

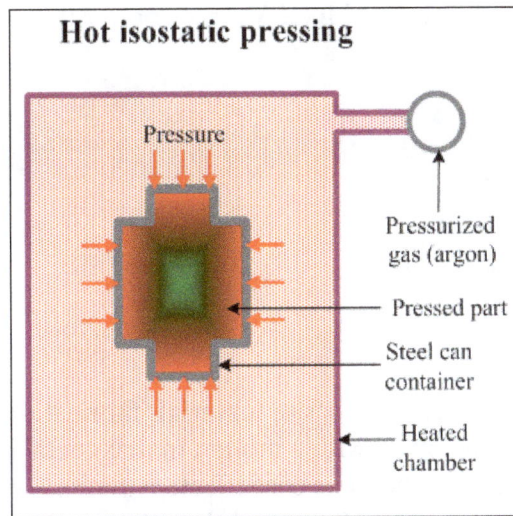

Hot Powder Extrusion Fabrication of Metal Matrix Composites

Hot Powder Extrusion Fabrication of Metal Matrix Composites – sintering under a pressure applied by an extruder at elevated temperature.

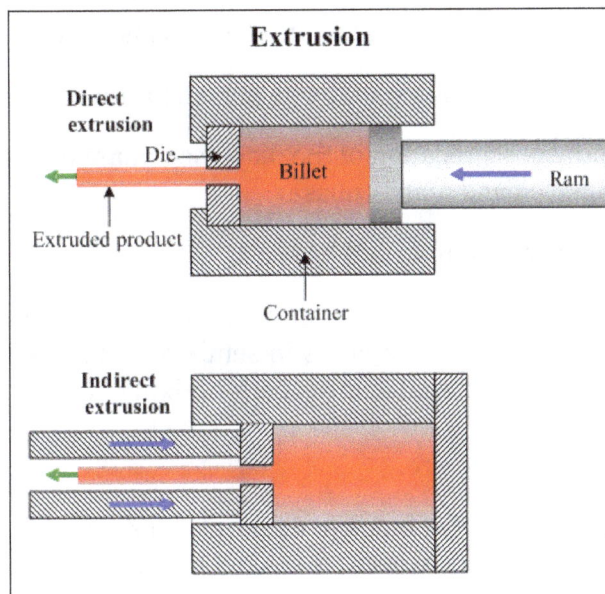

Extrusion

Direct extrusion

Die

Billet

Ram

Extruded product

Container

Indirect extrusion

Liquid State Processing

Liquid state fabrication of Metal Matrix Composites involves incorporation of dispersed phase into a molten matrix metal, followed by its Solidification. In order to provide high level of mechanical properties of the composite, good interfacial bonding (wetting) between the dispersed phase and the liquid matrix should be obtained.

Wetting improvement may be achieved by coating the dispersed phase particles (fibers). Proper coating not only reduces interfacial energy, but also prevents chemical interaction between the dispersed phase and the matrix.

The methods of liquid state fabrication of Metal Matrix Composites:

- Stir Casting
- Infiltration
- Gas Pressure Infiltration
- Squeeze Casting Infiltration
- Pressure Die Infiltration

Stir Casting

Stir Casting is a liquid state method of composite materials fabrication, in which a dispersed phase (ceramic particles, short fibers) is mixed with a molten matrix metal by means of mechanical stirring. Stir Casting is the simplest and the most cost effective method of liquid state fabrication. The liquid composite material is then cast by conventional casting methods and may also be processed by conventional Metal forming technologies.

Stir Casting is characterized by the following features:

- Content of dispersed phase is limited (usually not more than 30 vol. %).

- Distribution of dispersed phase throughout the matrix is not perfectly homogeneous:

  - There are local clouds (clusters) of the dispersed particles (fibers);

  - There may be gravity segregation of the dispersed phase due to a difference in the densities of the dispersed and matrix phase.

- The technology is relatively simple and low cost.

Distribution of dispersed phase may be improved if the matrix is in semi-solid condition. The method using stirring metal composite materials in semi-solid state is called Rheocasting. High viscosity of the semi-solid matrix material enables better mixing of the dispersed phase.

Infiltration

Infiltration is a liquid state method of composite materials fabrication, in which a preformed dispersed phase (ceramic particles, fibers, woven) is soaked in a molten matrix metal, which fills the space between the dispersed phase inclusions.

The motive force of an infiltration process may be either capillary force of the dispersed phase (spontaneous infiltration) or an external pressure (gaseous, mechanical, electromagnetic, centrifugal or ultrasonic) applied to the liquid matrix phase (forced infiltration). Infiltration is one of the methods of preparation of tungsten-copper composites.

The principal steps of the technology are as follows:

- Tungsten Powder preparation with average particle size of about 1-5 mkm.

- Optional step: Coating the powder with nickel. Total nickel content is about 0.04%.

- Mixing the tungsten powder with a polymer binder.

- Compacting the powder by a molding method (Metal injection molding, die pressing, iso-static pressing). Compaction should provide the predetermined porosity level (apparent density) of the tungsten structure.

- Solvent debinding.

- Sintering the green compact at 2200-2400 °F (1204-1315 °C) in Hydrogen atmosphere for 2 hrs.

- Placing the sintered part on a copper plate (powder) in the infiltration/sintering furnace.

- Infiltration of the sintered tungsten sceleton porous structure with copper at 2100-230 °F (110-126 °C) in either hydrogen atmosphere or vacuum for 1 hour.

Gas Pressure Infiltration

Gas Pressure Infiltration is a forced infiltration method of liquid phase fabrication of Metal Matrix Composites, using a pressurized gas for applying pressure on the molten metal and forcing it to penetrate into a preformed dispersed phase.

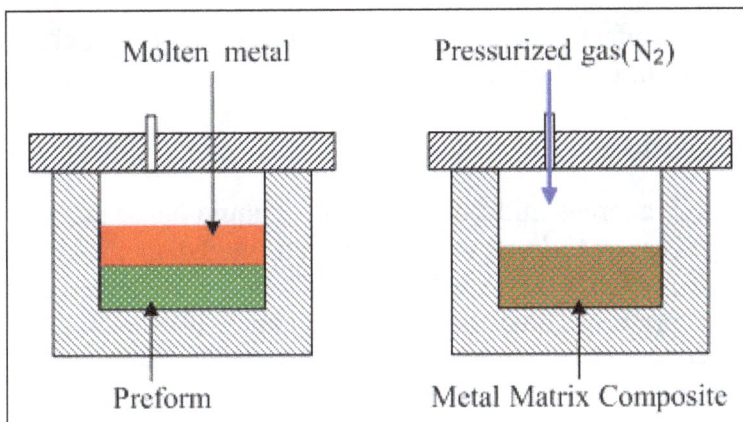

Gas Pressure Infiltration method is used for manufacturing large composite parts. The method allows using non-coated fibers due to short contact time of the fibers with the hot metal. In contrast to the methods using mechanical force, Gas Pressure Infiltration results in low damage of the fibers.

Squeeze Casting Infiltration

Squeeze Casting Infiltration is a forced infiltration method of liquid phase fabrication of Metal Matrix Composites, using a movable mold part (ram) for applying pressure on the molten metal and forcing it to penetrate into a performed dispersed phase, placed into the lower fixed mold part. Squeeze Casting Infiltration method is similar to the Squeeze casting technique used for metal alloys casting.

Squeeze Casting Infiltration process has the following steps:

- A preform of dispersed phase (particles, fibers) is placed into the lower fixed mold half.

- A molten metal in a predetermined amount is poured into the lower mold half.

- The upper movable mold half (ram) moves downwards and forces the liquid metal to infiltrate the preform.

- The infiltrated material solidifies under the pressure.

- The part is removed from the mold by means of the ejector pin.

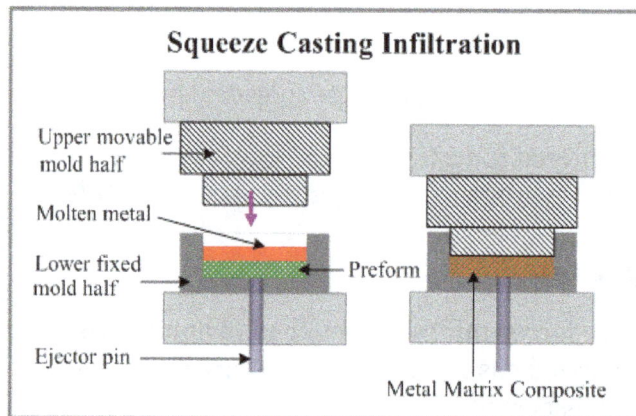

Squeeze Casting Infiltration

The method is used for manufacturing simple small parts (automotive engine pistons from aluminum alloy reinforced by alumina short fibers).

Pressure Die Infiltration

Pressure Die Infiltration is a forced infiltration method of liquid phase fabrication of Metal Matrix Composites, using a Die casting technology, when a preformed dispersed phase (particles, fibers) is placed into a die (mold) which is then filled with a molten metal entering the die through a sprue and penetrating into the preform under the pressure of a movable piston (plunger).

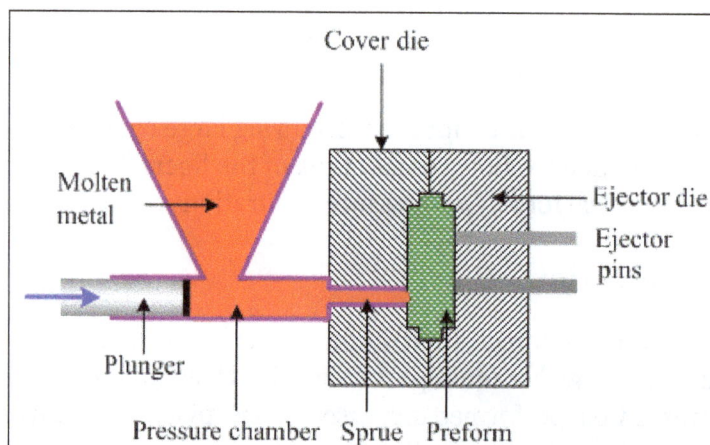

In-situ Processing

In situ fabrication of Metal Matrix Composite is a process, in which dispersed (reinforcing) phase is formed in the matrix as a result of precipitation from the melt during its cooling and Solidification.

Different types of Metal Matrix Composites may be prepared by in situ fabrication method:

- Particulate in situ MMC – Particulate composite reinforced by in situ synthesized dispersed phase in form of particles.

  Examples: Aluminum matrix reinforced by titanium boride (TiB_2) particles, magnesium matrix reinforced by Mg_2Si particles.

- Short-fiber reinforced in situ MMC – Short-fiber composite reinforced by in situ synthesized dispersed phase in form of short fibers or whiskers (single crystals grown in form of short fibers).

  Examples: Titanium matrix reinforced by titanium boride (TiB_2) whiskers, Aluminum matrix reinforced by titanium aluminide ($TiAl_3$) whiskers.

- Long-fiber reinforced in situ MMC – Long-fiber composite reinforced by in situ synthesized dispersed phase in form of continuous fibers.

  Example: Nickel-aluminum (NiAl) matrix reinforced by long continuous fibers of Mo (NiAl-9Mo alloy).

  Dispersed phases of in situ fabricated Metal Matrix Composites may consist of intermetallic compounds, carbides, borides, oxides, one of eutectic ingredients.

Advantages of In-situ Metal Matrix Composites

- In situ synthesized particles and fibers are smaller than those in materials with separate fabrication of dispersed phase (ex-situ MMCs). Fine particles provide better strengthening effect.

- In situ fabrication provides more homogeneous distribution of the dispersed phase particles.

- Bonding (adhesion) between the particles of in situ formed dispersed phase and the matrix is better than in ex-situ MMCs.

- Equipment and technologies for in situ fabrication of MMCs are less expensive.

Disadvantages of In-situ Metal Matrix Composites

- Choice of the dispersed phases is limited by thermodynamic ability of their precipitation in particular matrix.

- The size of dispersed phase particles is determined by solidification conditions.

Unidirectional solidification of a eutectic alloy (alloy of eutectic composition) may result in formation of eutectic structure, in which one of the components has a form of long continuous filaments.

Scheme of a device for unidirectional solidification of in situ Metal Matrix Composite is shown in the figure.

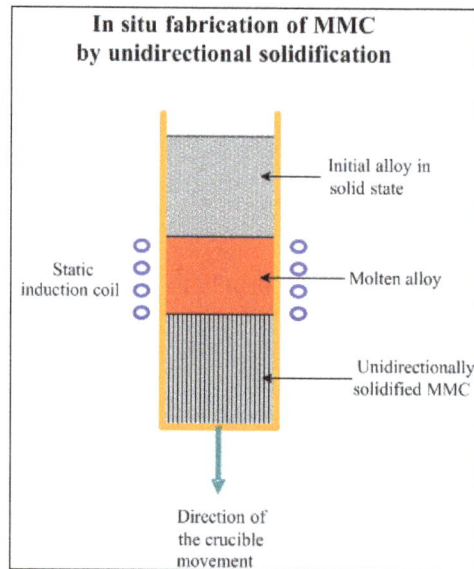

In situ fabrication of MMC by unidirectional solidification

Crucible with an eutectic alloy moves downwards (or alternatively the induction coil moves upwards). This movement results in remelting followed by resolidification of the alloy under controlled cooling conditions.

Value of heat transfer through the crucible bottom together with the crucible speed (v) and the power of the heating elements (induction coil) determine particular temperature gradient, which provides unidirectional solidification with flat solidification front.

The alloy acquires eutectic structure directed along the solidification direction with eutectic components in form of long mono-crystals (fibers). A distance between the fibers (d) is determined by the solidification speed (v) according to the formula:

$$d^2 \sim v.$$

Properties of Metal Matrix Composites

Physical and Mechanical Properties

Composite properties depend first and foremost on the nature of the composite. Composites are classified as complex, heterogeneous, and often anisotropic material systems. Their properties are affected by many variables including reinforcement form, volume fraction, geometry, distribution, matrix/reinforcement interface, void content, and manufacturing process. Hereby, representative mechanical and physical properties of MMCs at ambient temperature (i.e., room temperature) are presented for a broad range of mechanical engineering applications.

Mechanical Properties of MMCs

By reinforcing metals and alloys with continuous/discontinuous fibers, whiskers, and particles, higher strength and stiffness as well as better wear resistance can be achieved.

Figure compares the tensile curves for a typical reinforced material, and a conceptual fiber-reinforced MMC as well as a matrix material (i.e., a metal or alloy). Adding hard, brittle reinforcement particle to a ductile metal matrix can strengthen the metal significantly. When tensile stress is applied parallel to the fiber direction in a continuous fiber MMC, both fiber and matrix deform elastically. By keep applying higher stresses, the matrix undergoes plastic deformation while the fibers are still in the elastic regime. If the fibers possess some ductility, both matrix and fibers deform plastically; however, fibers break without experiencing plastic deformation. As the fibers fracture, the load is transferred back to the weaker matrix, and the composite fails immediately.

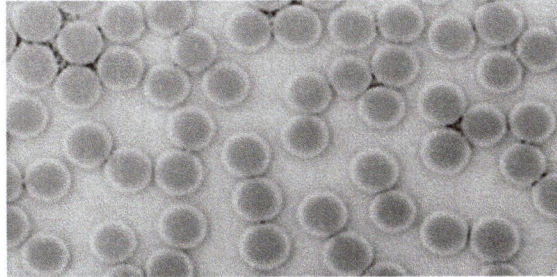

Structure of a unidirectional endless fiber reinforced aluminum composite material (transverse grinding): matrix: AA 1085, 52 vol% 15- mm Altex-fiber (Al_2O_3).

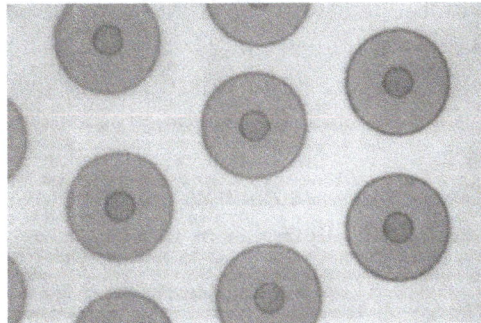

Structure of a titan matrix composite material of SiC monofilaments, World of MMC asses.

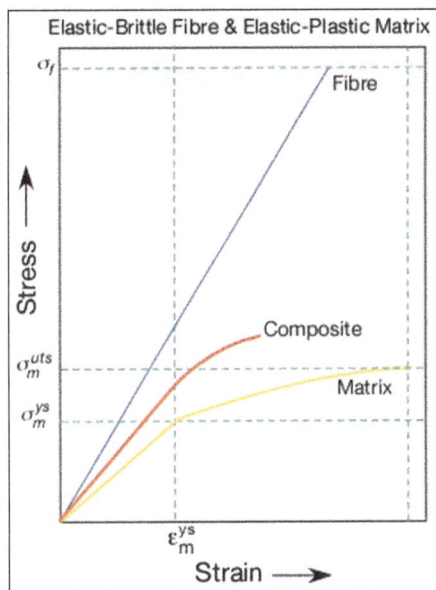

Schematic representation of tensile curves of matrix and reinforcement as compared with composite material.

The MMCs exhibit higher strength because they are able to transfer much of the loads to the strong reinforcement particles, reducing the stress carried by the matrix. The fiber–matrix interphase properties must be carefully tailored and maintained over the life of the composite to obtain the desirable behaviors. The strength of MMCs depends upon a much more complex manner on composite microstructure.

Modulus of Elasticity

The stiffness, which is proportional to the elastic modulus, of a MMC increases with introducing of reinforcement. Type and volume fraction of reinforcement are the main factors contributing the elastic behavior of a composite. For instance, in a unidirectional reinforced continuous fiber MMC, the longitudinal Young's modulus increases linearly as a function of the fiber volume fraction according to:

$$E_c = V_r E_r + V_m E_m$$

where E is the elastic modulus, V is the volume fraction and c, m, and r are indexes for the composite, matrix, and reinforcement, respectively.

In figure shows an example of modulus increase as a function of fiber volume fraction for an alumina fiber-reinforced aluminum lithium alloy matrix composite. The increase in the longitudinal Young's modulus is in agreement with the rule-of-mixtures value, whereas the modulus increase in a direction transverse to the fibers is much lower. Particle reinforcement also results in an increase in the modulus of the composite; the increase, however, is much less than that predicted by the rule-of-mixtures. This is understandable inasmuch as the rule of mixtures is valid only for continuous fiber reinforcement. The relatively high cost of many continuous reinforcing fibers used in MMCs has limited the application of these materials.

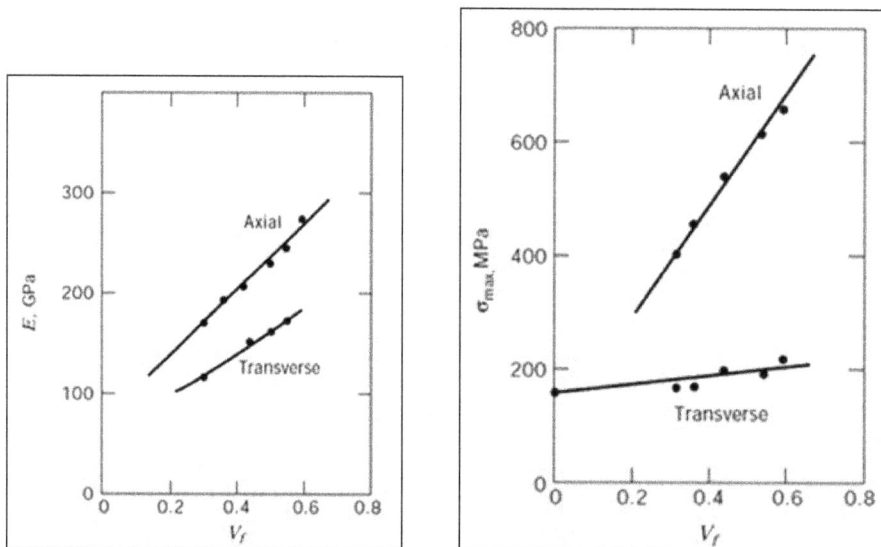

Modulus increase as a function of fiber volume fraction Vf for alumina fiber-reinforced aluminum lithium alloy matrix for (a) E (elastic modulus), and (b) σ_{max}.

The most widely used MMCs are reinforced with discontinuous fibers or particles. For discontinuously reinforced MMCs,

$$E_c = \frac{E_m 1 + 2sqV}{1 - qV_r}$$

where,

$$q = \frac{\left(E_r / E_m - 1\right)}{\left(E_r / E_m\right) + 2_s}$$

Where s is the particle aspect ratio.

Strength

The introduction of reinforcements to matrix alloy generally enhances both yield and ultimate tensile strength. The strength of a fiber-reinforced composite is determined by fracture processes, themselves governed by a combination of microstructural phenomena and features. These include plastic deformation of the matrix, the presence of brittle phases in the matrix, the strength of the interface, the distribution of flaws in the reinforcement, and the distribution of the reinforcement within the composite. Consequently, predicting the strength of the composite from that of its constituent phases is generally difficult. In some cases, the improvements are dramatic. The greatest increases in strength and modulus are achieved with continuous fibers. Figure shows Ashby plot, strength vs. density for different classes of materials including composites.

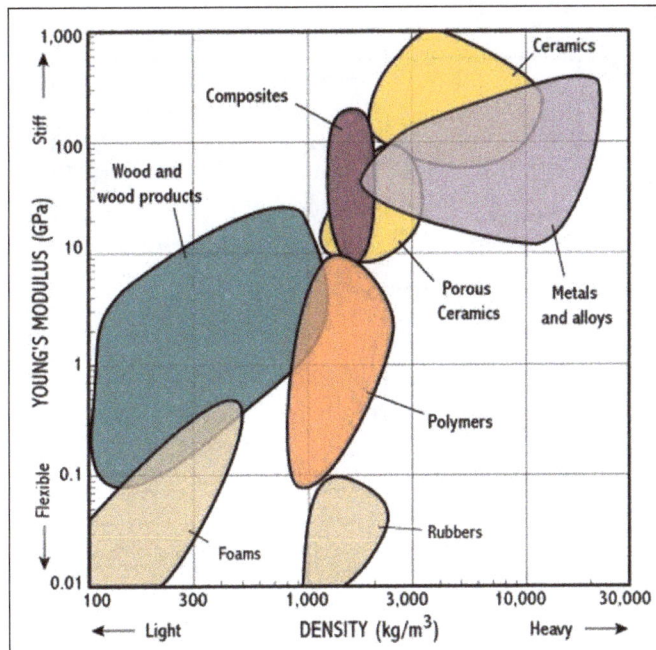

Young's modulus versus density for different classes of materials.

Elongation

Elongation of the MMCs is generally less than matrix alloy. For instance, the elongation of a 6061-T6 aluminum alloy is about 12%; by adding 15 vol% Al_2O_3 particles to this alloy, the elongation drops to 5.4%. This is mainly due to the fact that the composite failure is associated with particle cracking and void formation in the matrix within the particle clusters. Also, in MMCs, the strain in the matrix is constrained with surrounding reinforcement phases (i.e., particles, fibers). Size, type, distribution of particles, matrix microstructure, and nature of matrix-reinforcement interface are the main contributing factors in MMCs' ductility.

Fatigue

Fatigue life of composites is significantly increased by adding the reinforcements. The range of the stress intensity factor, ΔK, for Al–SiCp is between 2 and 4 MPa m$^{1/2}$ which is twice that of unreinforced aluminum (1–2 MPa m$^{1/2}$). The fatigue resistance of long-fiber reinforced MMCs is larger than that of unreinforced metals when loaded in tension along the fiber axis. Crack deflection and reduction in slip band formation are the main contributing factors to the superior fatigue properties of MMCs. In terms of general S-N curve behavior, a composite such as a silicon carbide particle-reinforced aluminum-based MMC shows an improved fatigue behavior compared to the reinforced alloy. Such an improvement in stress controlled cyclic loading or high cyclic fatigue is attributed to the higher stiffness of the composite. However, the fatigue behavior of the composite, evaluated in terms of strain amplitude vs. cycles or low cycle fatigue, was found to be inferior to that of the unreinforced alloy. This was attributed to the lower ductility of the composite compared to the unreinforced alloy. At elevated temperatures, particularly close to the aging temperature for age hardenable alloys, the matrix strength decreases resulting in a decrease in fatigue strength.

Fracture Toughness

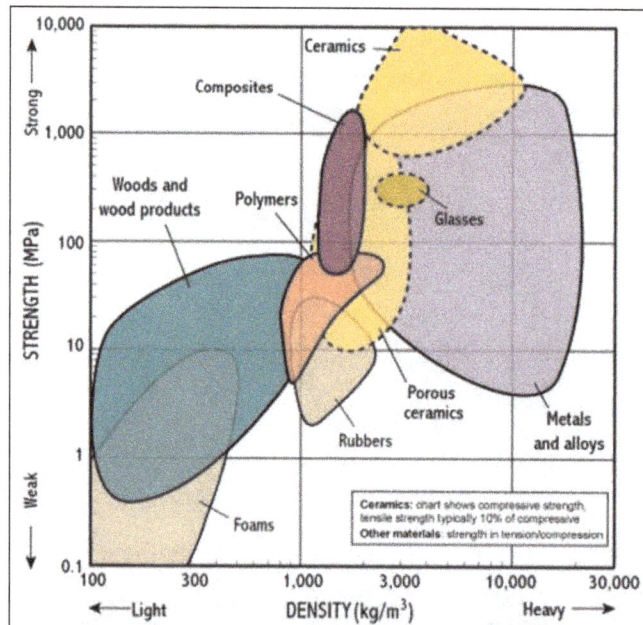

Strength versus density for different classes of material.

Fracture toughness of particulate composites generally decreases with increase in particle size at a given volume fraction and inter-particle spacing. Fracture toughness decreases with increasing tensile strength for cast composites, whereas it increases by 70–100% for forged ones over that of cast ones. This increase in toughness with increasing amount of particulates is related to blunting of the original crack tip caused by the void nucleated over the particles. While the crack tip radius increases, the stress concentration at crack tip decreases resulting in combined increase in facture toughness values. Unidirectional fibers reinforcement can lead to easy crack initiation and propagation vis-à-vis the unreinforced alloy matrix. Braiding of fibers can make the crack propagation toughness increase largely due to extensive matrix deformation, fiber bundle deboning, and pull out.

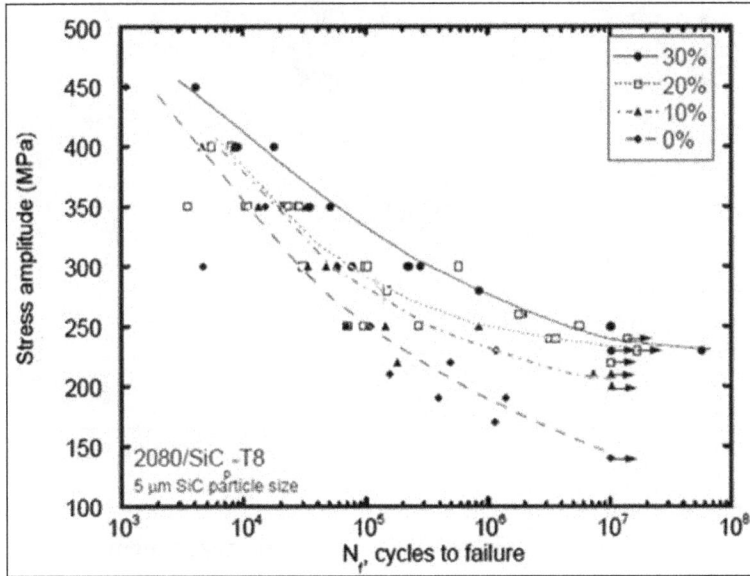

Stress versus cycles (S–N) behavior of a particle reinforced MMC. With increasing volume fraction of particles the fatigue strength of the composite increases.

Creep

It has been reported that the creep resistance of $Al_9Si_3Cu/25$ vol% Al_2O_3 composites at 400 °C increased by 100% compared to that of unreinforced alloys. The dependence of creep rate on both stress level and temperature is more in the composites compared to the matrix metal/alloy. The whisker-reinforced composites are more creep resistance than particulate composites.

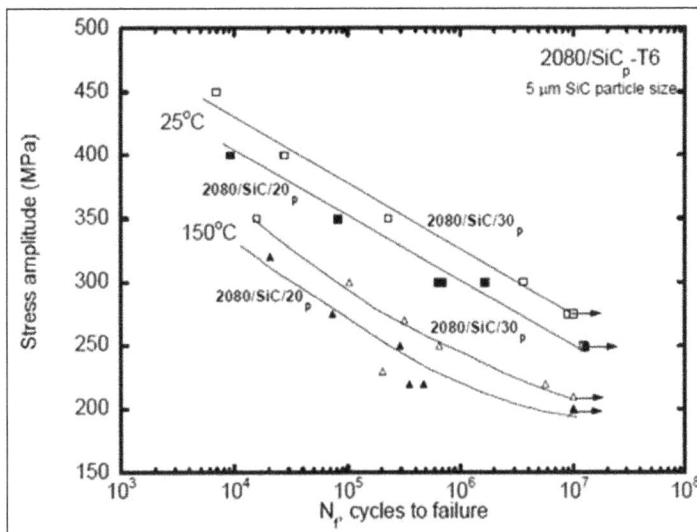

Elevated-temperature fatigue behavior of a particle reinforced MMC. With increasing temperature, the matrix strength decreases resulting in decreased fatigue resistance (aging temperature was 175 °C).

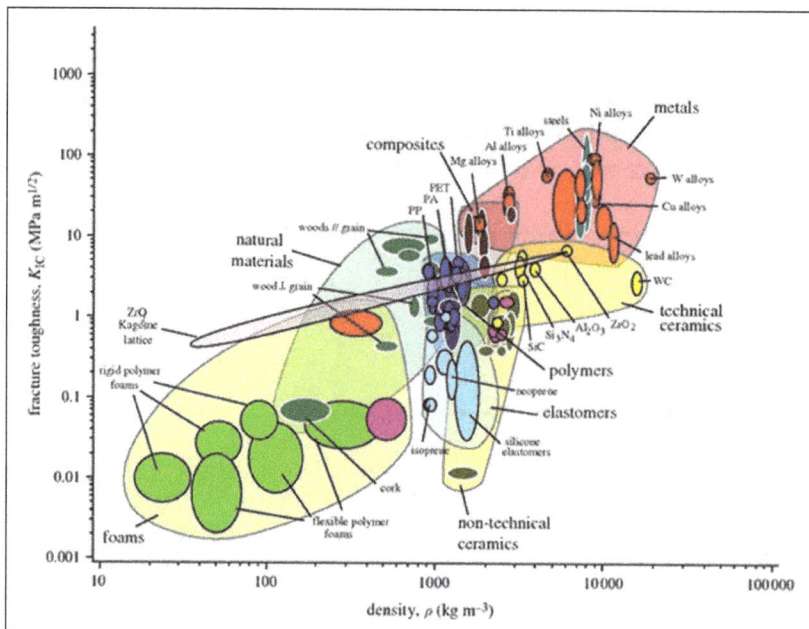

Fracture toughness vs. Young's modulus for different materials including composites.

The addition of short-fiber reinforcement enhances the creep strength due to effective load transfer to the fibers. Whisker or particle reinforcement also results in significant creep strengthening over the unreinforced alloy. Anomalously high values of the activation energy, Q, and stress exponent, n, have been reported in MMCs. Nardone and Strife rationalized this by proposing the concept of a threshold stress sr, for creep deformation, originally used to explain the high values for Q and n in dispersion-strengthened alloys. The physical explanation for the threshold stress in the discontinuously reinforced composite system can be attributed to a variety of reasons: (1) Orowan bowing between particles, (2) back-stress associated with dislocation climb, (3) attractive force between dislocations and particles, resulting from relaxation of the strain field of dislocations at the particle/matrix interface.

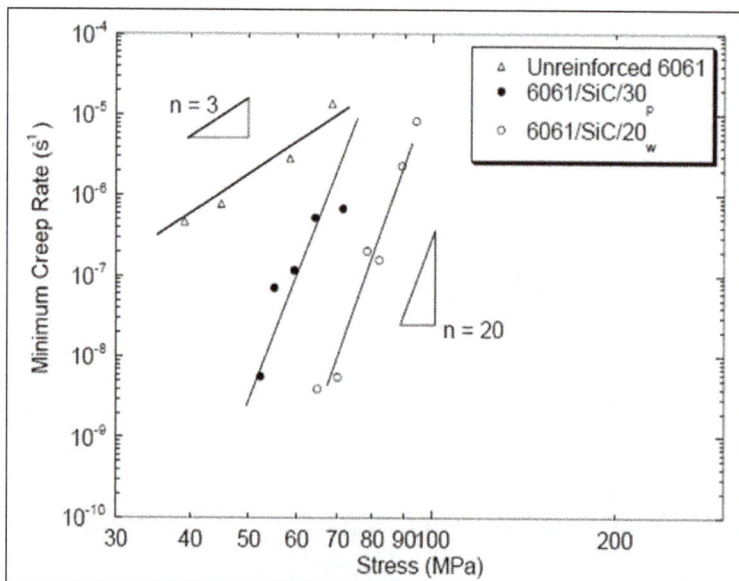

Steady-state creep rate as a function of applied stress for SiC-particle and SiC whisker-reinforced Al–matrix composites.

Physical Properties of MMCs

Density, CTE, and thermal conductivity are among most commonly discussed physical properties of composites. Composites are considered multi-functional materials; among the family of composites, there are an increasing number of material systems than combine superior physical properties including high thermal conductivity along with low density and adaptable CTE as well as excellent mechanical properties.

Physical properties of matrix can be significantly altered by addition of a reinforcement and chiefly depend on the reinforcement distribution. A good example is aluminum–silicon carbide composites, for which the presence of the ceramic increases, substantially, the elastic modulus of the metal without greatly affecting its density. Elastic moduli for 6061 aluminum–matrix composites reinforced with discrete silicon carbide particles or whiskers have been calculated by using the rule of mixtures for the same matrix reinforced with two types of commercial continuous silicon carbide fibers. As a result, several general facts become apparent; first, modulus improvements are significant, even with equiaxed silicon carbide particles, which are far less expensive than fibers or whiskers. However, the level of improvement depends on the shape and alignment of the silicon carbide. Also, it depends on the processing of the reinforcement: for the same reinforcement shape (continuous fibers), microcrystalline polycarbosilane-derived SiC fibers yield much lower improvements than do crystalline polycarbosilane SiC fibers. These features, which influence reinforcement shape, orientation, and processing of modules, are quite general; they are also observed, for example, in MMC reinforced with aluminum oxide or carbon.

Strengthening Methods in MMCs

Dislocation Strengthening

Dislocations' density is higher in the composite matrix than in unreinforced metal with the same history. In a composite, dislocations can be generated by (1) straining in response to an applied load, (2) straining to relax residual thermal stresses caused by CTE mismatches between matrix and reinforcement. The increase in dislocations' density can be written as,

$$\Delta\sigma = 12 \frac{\Delta\alpha\Delta Tf}{bd}$$

where $\Delta\rho$ is the increase in dislocation density, $\Delta\alpha$ is the CTE mismatch, ΔT is the temperature difference, b is Burgers vector, f is the reinforcement volume fraction, and d is particle size. From this equation, the change in matrix yield strength can be estimated:

$$\Delta\sigma = Gb\sqrt{\Delta\rho}$$

Therefore the dislocation density and hence matrix strengthening increase with decreasing reinforcement particle size and increasing reinforcement volume fraction.

Orowan Strengthening

This mechanism strengthens the composite by making subsequent dislocation motion more difficult. Although important in dispersion-strengthened systems, Orowan mechanism does not effectively strengthen other MMCs because in other MMCs the reinforcement phases (particles) are too large and to too far to be considered as effective obstacle to the motion of dislocations. However, Orowan strengthening mechanism can be significant in discontinuously reinforced aluminum composites where the particles are finer and closer together and causes increase in the strength of the MMC.

Grain Refinement Strengthening

The MMCs usually possess much finer grain size than that of unreinforced matrix metal. Hall–Petch relationship implies that the beginning of plastic flow in the matrix depends on the local magnification of stress at grain boundaries resulting from dislocation pile-ups. A small grain size results in fewer dislocations in the pile-ups, and hence the applied stress must be higher to cause yielding.

Fabrication of MMCs

For a given MMC application, the processing (fabrication) method is a key factor that determines both cost and properties. The main challenges in fabricating MMCs include (1) finding a cost effective method to distribute the reinforcement phase in the reinforcement phase in the desired configuration, (2) achieving a strong bond between the matrix and the reinforcement to allow an effective load transfer between phases without any failure. The MMCs can be produced by a variety of fabrication techniques including solid state processes, liquid-state processes and deposition processes.

In solid-state processes, the most spread method is powder metallurgy; it is usually used for high melting point matrixes and avoids segregation effects and brittle reaction product formation prone to occur in liquid state processes. This method permits to obtain discontinuously particle reinforced aluminum–matrix composites with the highest mechanical properties. The aluminum– matrix composites are used for military applications but remain limited for large-scale productions. In liquid-state processes, one can distinguish the infiltration processes where the reinforcements form a preform which is infiltrated by the alloy melt (1) with pressure applied by a piston (squeeze-casting) or by an inert gas (gas pressure infiltration) and (2) without pressure. MMC fabrication under no pressure can be subdivide to (1) reactive infiltration, in which the wetting between reinforcement and melt is obtained by reactive atmosphere, elevated temperature, alloy modification or reinforcement coating and (2) the dispersion processes, such as stir-casting, where the reinforcements are particles stirred into the liquid alloy. Process parameters and alloys are to be adjusted to avoid reaction with particles. In deposition processes, droplets of molten metal are sprayed together with the reinforcing phase and collected on a substrate where the metal solidification is completed. This technique has the main advantage that the matrix microstructure exhibits very fine grain sizes and low segregation. This technique however has several drawbacks: the technique can only be used with discontinuous reinforcements, the costs are high, and the products are limited to the simple shapes that by obtained by extrusion, rolling or forging.

Another widely used fabrication method comprises spray processes. In this case, a molten metal stream is fragmented by means of a high-speed cold inert-gas jet passing through a spray gun, and dispersoid powders are simultaneously injected. A stream of molten droplets and dispersoid powders is directed toward a collector substrate where droplets recombine and solidify to form a high-density deposit.

Depending on the process, the desired microstructure, and the desired part, MMCs can be produced to net or near-net shape; or alternatively they can be produced as billet or ingot material for secondary shaping and processing. Figure, schematically, shows different techniques for making MMCs.

Applications of Metal Matrix Composites

The combined attributes of MMCs, together with the costs of fabrication, vary widely with the nature of the material, the processing methods, and the quality of the product. In engineering, the type of composite used and its application vary significantly, as do the attributes that drive the choice of MMCs in design. For example, high specific modulus, low cost, and high weld-ability of extruded aluminum oxide particle-reinforced aluminum are the properties desirable for bicycle frames. High wear resistance, low weight, low cost, improved high-temperature properties, and the possibility for incorporation in a larger part of unreinforced aluminum are the considerations for design of diesel engine pistons. MMCs are being used in various applications including:

- Aerospace
- Transportation
- Electronics
- Electric power transmission
- Recreational products and sporting goods
- Wear-resistance materials

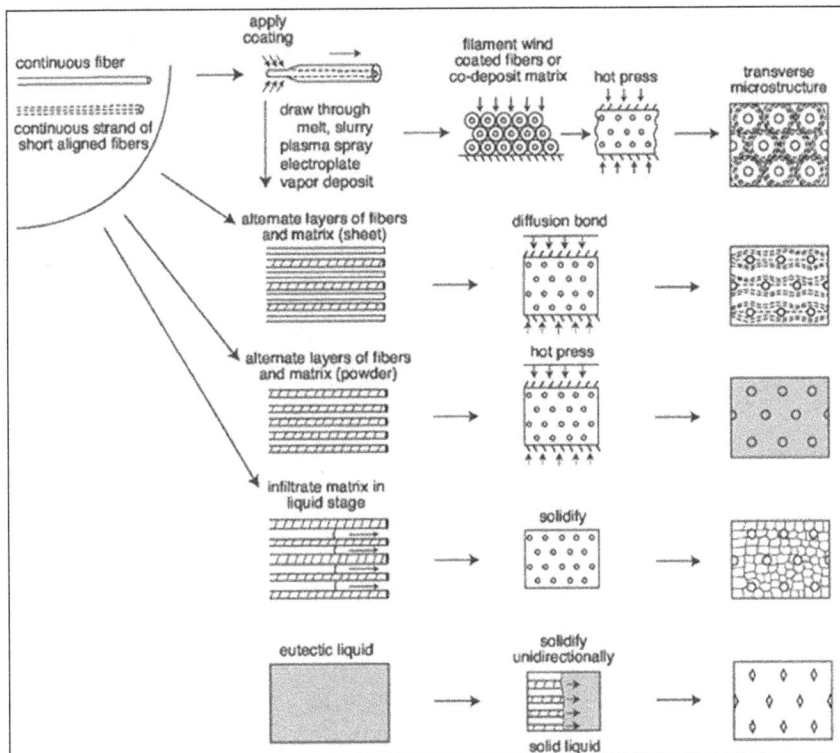

Methods used to make MMCs.

In figure shows the usage of MMCs per year, broken down by market segment (i.e., automotive, aerospace, consumer products). the ground transportation industry is, by far, the major market segment for MMCs. Applications in this segment include drive shafts, engine components, and brake components. The main constraint in the transportation sector is that it is an extremely cost-sensitive sector. Thus, reducing the manufacturing costs of the MMC components will greatly aid in the replacement of conventional parts.

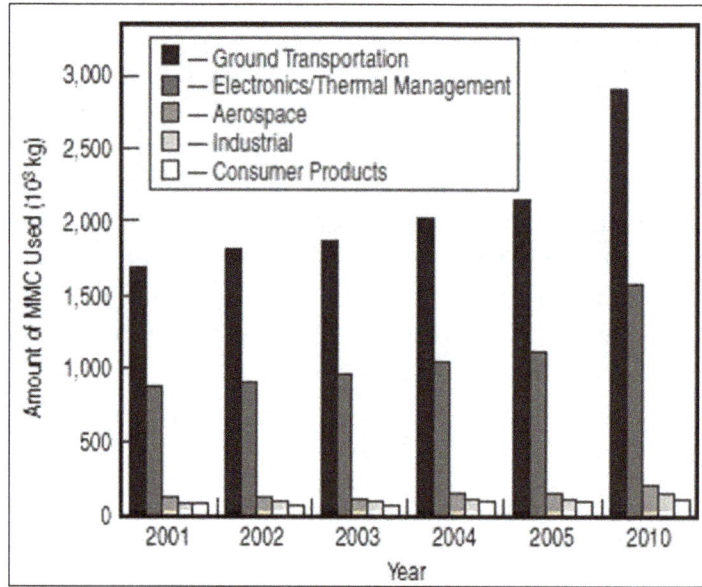

The use of MMCs in different market sectors.

The application of MMCs in aerospace industries is due to the fact that materials with enhanced specific stiffness and strength can significantly increase the performance of the aircraft. The MMCs have been used largely in military and commercial aircrafts. In the F-16 aircraft, for instance, aluminum access doors have been replaced with SiC-particle reinforced MMCs resulted in increased fatigue life. Continuous fiber-reinforced MMCs have also been used in military applications, due to high specific strength, stiffness, and fatigue resistance. SiC monofilament reinforced Ti–matrix composites have been used as nozzle actuator controls for the F119 engine in F-16. The MMC replaced a heavier Inconel 718 in the actuator links and stainless steel in the piston rods.

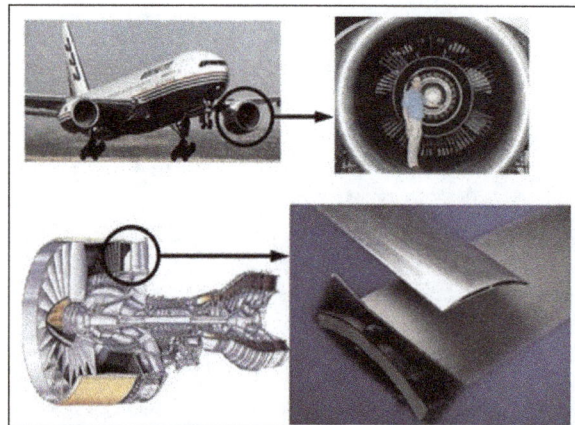

Application of a SiC-particle reinforced aluminum-based MMC in the fan-exit guide vane of a Pratt & Whitney engine on a Boeing 777.

MMCs are being used in commercial aircraft as well. Figure shows an application on the MMCs in the fan-exit guide vane of a Pratt & Whitney engine on a Boing 777. The MMC replaced a Carbon/Epoxy composite that had problems with foreign object damage (FOD). Boeing 787 is the first commercial jet transport to be manufactured out of predominantly composite materials. Traditional metallic materials are characterized by the isotropic nature of their material properties (no preferred directions in terms of tensile strength) and the fact that the predominant manufacturing method is based on material removal (milling). Composite materials, on the other hand, frequently are fibrous and anisotropic. Moreover, the predominant manufacturing methods are based on material deposition rather than removal.

Boeing 787 makes greater use of composite materials in its airframe and primary structure than any previous Boeing commercial airplane. Undertaking the design process without preconceived ideas enabled Boeing engineers to specify the optimum material for specific applications throughout the airframe. The result is an airframe comprising nearly half carbon fiber reinforced plastic and other composites. This approach offers weight savings on average of 20% compared to more conventional aluminum designs.

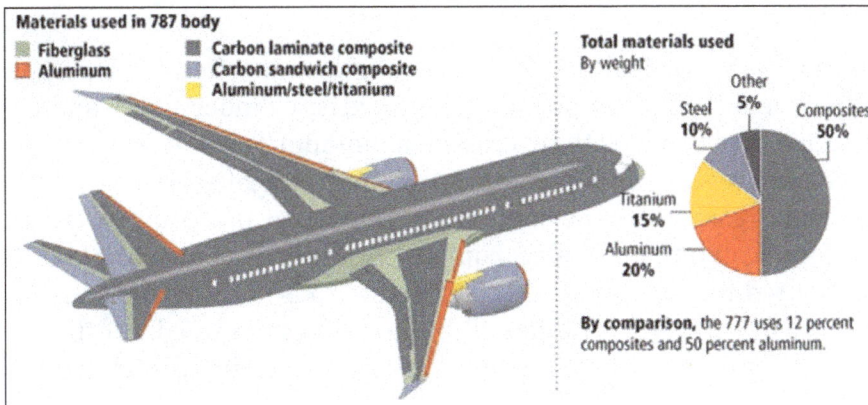

The percentages of materials used in Boing 787 aircraft.

MMCs have been used in a variety of automotive applications as well. An early MMC application in an automotive engine was a hybrid particulate-reinforced aluminum–matrix composite used as a cylinder liner in the Honda Prelude. The composite consisted of an Al–Si matrix with 12% Al_2O_3 for wear resistance and 9% carbon lubricity. The composite was integrally cast with the engine block, had improved cooling efficiency, and exhibited improved wear and a 50% weight savings over cast iron, without increasing the engine package size.

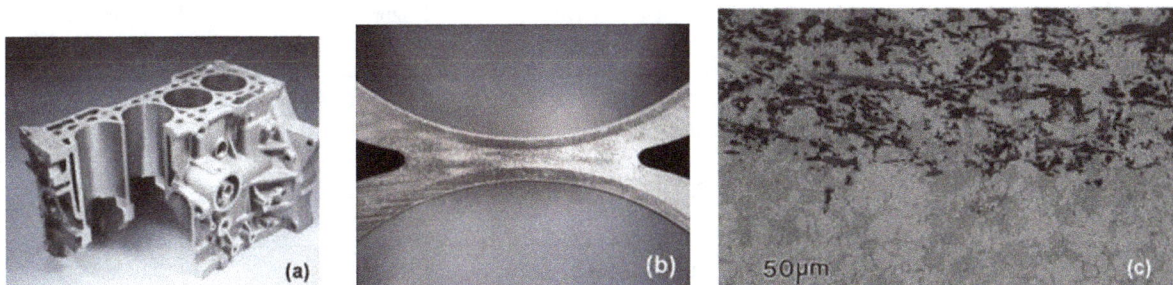

Hybrid particulate-reinforced aluminum-matrix composite used as a cylinder liner in the Honda Prelude; (a) Prelude engine block, (b) magnified view of cylinder liner, and (c) microstructure of composite showing carbon short fibers (black) and Al_2O_3 fibers (dark gray).

The automotive market is a high volume and a high technology market, but costs should be as low as possible. However, there are still a lot of reasons to consider the use of light aluminum composites:

- Reduction of the weight of engine parts;

- Increase of the operation temperature of engines;

- Improvement of the tribological properties of moving and contacting;

- Components (wear resistance, lubrication);

- Increase of stiffness and strength,

- Matching coefficient of thermal expansion (i.e., steel or cast iron in connection with aluminum alloys);

- The use of related manufacturing techniques (especially for discontinuously reinforced aluminum alloys).

A list of typical MMCs, in particular aluminum–matrix composites, for car applications is given in Table. Other current applications are piston parts, cylinder liners and connecting rods. The first high volume application is the successful aluminum Toyota-piston ring, reinforced with short Saffil fibers and produced by squeeze casting. Both weight saving and increasing wear resistance are the main reasons for the success. An important potential replacement of steel by SiC-particle-reinforced aluminum–matrix composites is in the connecting rod. Near-net-shape sinter-forging was used to fabricate MMCs connecting rods with tensile and fatigue properties comparable to those of extruded materials. Table compares the weight of the MMC connecting rod versus that of the steel rod. A 57% weight savings was achieved with the MMC rod, with a moderate increase in cost.

Table: Aluminum–matrix composites for car applications.

Material	Application	Improved Property	Feature	Manufacturer
Al-shore fiber	Piston ring	Abrasion resistance performance, lower cost	High-temperature engine	Toyota
Al-shore fiber	Piston combustion bowl	High-temperature performance	Improved durability	Consortium
Al-shore fiber	Selective reinforcement of motor block	High-temperature performance	Improved durability	Peugeot
Al-shore fiber	Cylinder liner	Improve stiffness, wear resistance, better heat	Improved durability	Honda
Al-SiC particles	Connecting rod	Specific strength, specific stiffness	Higher performance	Nissan
Al-Al$_2$O$_3$ fiber	Connecting rod	Specific strength, specific stiffness	Higher performance	Chrysler

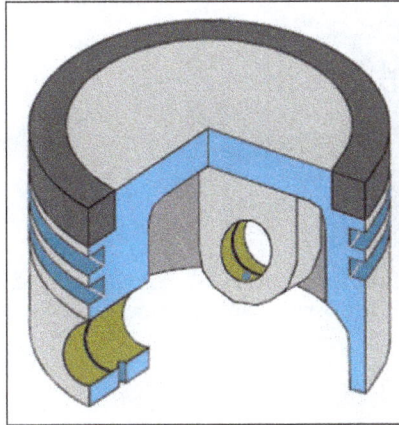

Selectively Reinforced Automotive Piston.

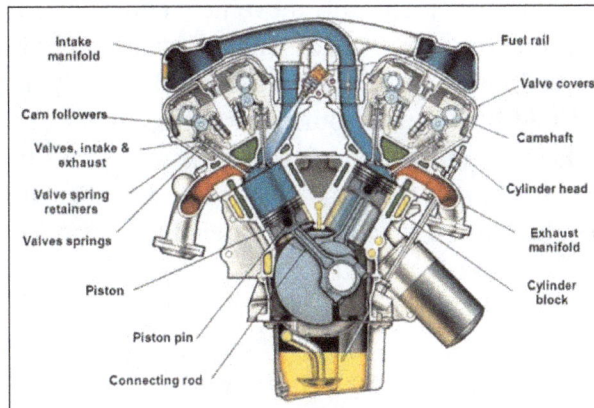

Cross-section of a passenger car engine showing the location of the connecting rod.

In figures the particulate MMCs for use in brake drums and brake rotors, as a replacement for cast iron. The high wear resistance and thermal conductivity coupled with 50–60% weight savings make the MMCs quite attractive for this application. Figure shows another example of the MMCs in automotive applications which is a 6061/Al_2O_3/20 p composite used as driveshaft in the Corvette. The composite exhibits a 36% increase in specific modulus over steel.

Table: Mass comparison of connecting rod for steel and an MMC.

	2080/SiC/20	Steel
Pine weight (gr)	65.2	144.7
Crank weight (gr)	184.0	437.7
Total weight (gr)	249.2	582.4

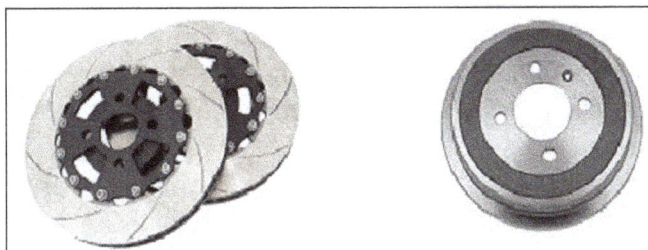

Brake drums and brake rotors made of MMC.

MMC made driveshaft in the Corvette with 36% increase in specific modulus over steel.

Railroad Brakes

The driving force for lightweight railway vehicles has also prompted the use of high-performance MMCs. A conventional brake system for a railway vehicle consists of four brake disks, calipers, hand brake, and electromagnetic track brake. This makes up about 20% of the total weight of the bogey. A particulate-reinforced aluminum–matrix composite (AlSi7Mg + SiC particulates, supplied by Duralcan) is fabricated by a multi-pouring process, where alternating layers of the unreinforced alloy and MMC is cast in successive layers. This contributed to reduced cost by using less MMC and placing the composite in the strategically important region (i.e., in contact with the wearing surface). In this application, the weight is reduced from 115 kg for a spheroidal graphite iron disk to 65 kg for the MMC disk, a weight savings of 43%. The steel brake shows significant cracking, while the MMC brake is in relatively good condition.

Electronic and Communication Applications

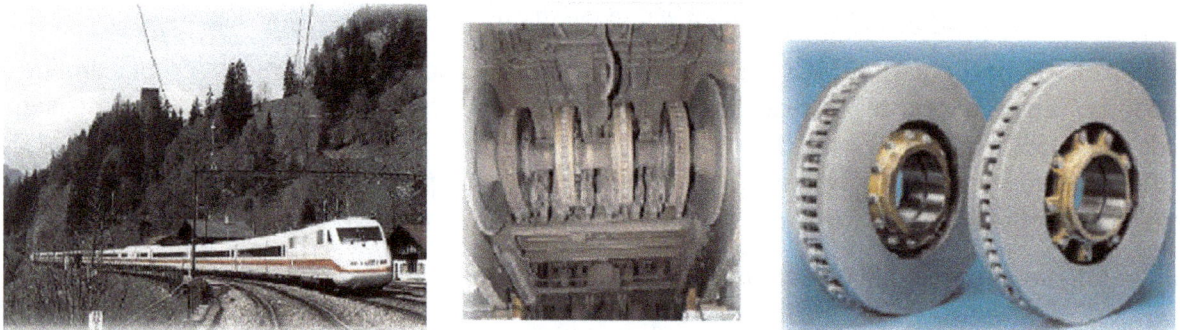

New generation advanced integrated circuits are generating more heat than previous types. Therefore, the dissipation of heat becomes a major concern in electrical applications. Indeed, thermal fatigue may occur due to a small mismatch of the CTE between the silicon substrate and the heat sink (normally molybdenum). This problem can be solved by using the MMCs with exactly matching coefficients (i.e., aluminum with boron or graphite fibers and aluminum with SiC particles). Besides a low CTE and a high thermal conductivity, the aluminum-based MMCs also have a low density and a high elastic modulus. Figure shows cross-section of an electrical conductor for power transmission including alumina fiber (Nextel 610)/Al core strands, Al–Zr alloy outer strands. The core consists of individual wires made from a continuously reinforced aluminum MMC produced by 3 M. The MMC core supports the load for the 54 aluminum wires and also carries a significant current, unlike competing steel cores. Hermetic package materials are developed to protect electronic circuits from moisture and other environmental hazards. These packages have often glass-to-metal seals. Therefore, materials with an 'adjustable' CTE are required. Al-based MMCs are fulfilling this condition, as the CTE is depending upon the volume fraction of the fibers or particles.

Brake rotors for German high speed train ICE-2 made from a particulate-reinforced aluminum alloy (AlSi7Mg + SiC particulates, supplied by Duralcan), developed by Knorr Bremse AG. Compared to conventional parts made out of cast iron with 115 kg/piece the 65 kg of the MMC rotor offers an attractive weight saving potential: each wagon has 8 brake rotors and in combination with the reduction of un-damped masses a highly efficient component is realized with MMC.

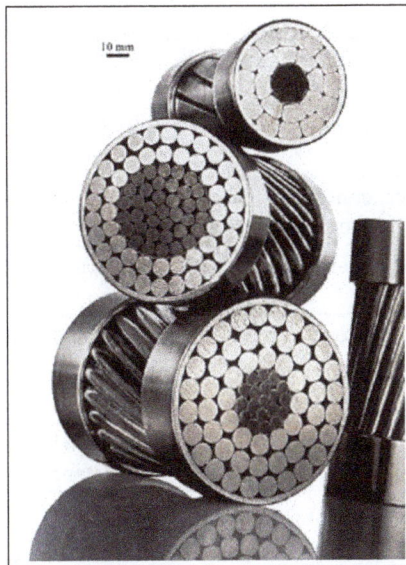

Aluminum conductor composite reinforced (ACCR).

Sports and Leisure Market Applications

The already well known advantages of Al-based composites are leading to several applications in various leisure and sporting goods. Typical applications are fishing rods, tennis and squash rackets, bicycle frames, golf club heads, track and field shoes, and cheetah flex-foot.

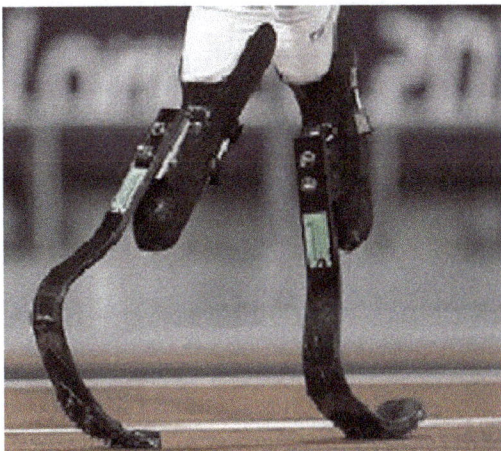

MMCs in track shoes and Cheetah Flex-Foot Carbon Fiber Reinforced composite.

Table: Potential and realistic technical applications of metal–matrix composites.

Composite	Components	Advantages
Aluminum–silicon carbide (particle)	Piston Brake rotor, caliper, line	Reduced weight, high strength and wear resistance High wear resistance, reduced weight.
	Propeller shaft	Reduced weight, high specific stiffness.
Aluminum–silicon carbide (whiskers)	Connecting rod	Reduced reciprocating mass, high specific strength and stiffness, low coefficient of thermal expansion.
	Sprockets, pulleys, and covers	Reduced weight, high strength and stiffness.
Aluminum–aluminum oxide (short fibers)	Piston ring	Wear resistance, high running temperature.
	Piston crown (combustion bowl)	Reduced reciprocating mass, high creep and fatigue resistance.
Aluminum–aluminum oxide (long fibers)	Connecting rod	Reduced reciprocating mass, improved strength and stiffness.
Copper–graphite	Electrical contact strips, electronics packaging, bearings	Low friction and wear, low coefficient of thermal expansion.
Aluminum–graphite	Cylinder, liner platon, bearings	Call resistance, reduced friction, wear and weight.
Aluminum–titanium carbide (particle)	Piston, connecting rod	Reduced weight and wear.
Aluminum–fiber flax	Piston	Reduced weight and wear.
Aluminum–aluminum oxide fibers–carbon fibers	Engine block	Reduced weight, improved strength and wear resistance.
Supper alloy-based composite (Ni–Ni$_3$Nb)	Turbine blades	Fatigue resistance, impact strength, temperature resistance.
Cu–Nb, Cu–Nb$_3$Sn	Super conductor	Superconducting, mechanical strength, ductility.
Cu–W	Spot welding electrodes	Burn-up resistance.

References

- Metal-matrix-composites: machinedesign.com, Retrieved 4 March, 2019

- Metal-Matrix-Composites: researchgate.net, Retrieved 11 May, 2019

- Solid-state-fabrication-of-metal-matrix-composites: substech.com, Retrieved 17 January, 2019

- Liquid-state-fabrication-of-metal-matrix-composites: substech.com, Retrieved 7 April, 2019

- In-situ-fabrication-of-metal-matrix-composites: substech.com, Retrieved 28 February, 2019

Ceramic Matrix Composites

The composite materials which are made up of ceramic fibers which are embedded in a ceramic matrix are known as ceramic matrix composites. Processing of ceramic matrix composites can be done using a variety of techniques such as chemical vapor infiltration and liquid phase infiltration. The diverse aspects of ceramic matrix composites have been thoroughly discussed in this chapter.

Ceramic matrix composites (CMCs) are being developed to take advantage of the high-temperature properties of ceramics while overcoming the low fracture toughness of monolithic ceramics. Toughening mechanisms, such as matrix cracking, crack deflection, interface debonding, crack-wake bridging, and fiber pullout, are being incorporated in CMCs to reduce the tendency for catastrophic failure found in monolithic ceramics. Ceramics reinforced with particulate, whiskers, and continuous fibers exhibit varying aspects of these toughening mechanisms; however, reinforcement with continuous fibers offers the greatest improvements in toughness. Composites with carbide, oxide, glass, and carbon matrices are being utilized in the development of CMCs. In the case of carbide, oxide, and glass matrix CMCs, the matrix exhibits excellent high-temperature corrosion resistance so that a goal of the composite development is to not detract from this preexisting property. This is not the case for carbon matrix composites, which frequently need coatings to provide adequate corrosion protection.

Composite materials are chemically and microstructurally heterogeneous, consisting of matrix, matrix–reinforcement interface, and reinforcement constituents. The corrosion behavior of each of these constituents will likely not be equal whether evaluated individually or within the composite. The composite corrosion resistance may be more complex than each constituent because of interactions between corrosion reaction products and the composite constituents. An example of this interaction would be a reaction product from the corrosion of the matrix that protects a less corrosion-resistant reinforcement or reinforcement– matrix interface. However, there is far more data on the corrosion behavior of the constituents as monolithic materials than there are on the composites made from these constituents. It is important to note that the constituents of a composite may differ slightly or substantially in chemical composition, crystal structure, and microstructure from their monolithic equivalent. Examples of these differences are the SiC fibers and matrix produced from polycarbosilane or its variants that may contain considerable oxygen or may be nanocrystals embedded in an amorphous matrix. There is considerable value in evaluating the corrosion resistance of the monolithic equivalent of the composite constituents because these generally represent the baseline corrosion behavior. An effort will be made to identify where significant differences are likely to occur between the composite constituents and the monolithic equivalent.

High-temperature composites are being developed to operate in a variety of environments containing alkali elements, mildly oxidizing mixed gases such as $He+O_2$, highly oxidizing environments, and H_2. Not all ceramic composites are being considered for each of these environments, so the corrosion data are not available for each ceramic composite in every environment.

Silicon Carbide Matrix Composites

Corrosion Reactions for Silicon Carbide

Silicon carbide will chemically react with O_2, H_2, and H_2O according to the following reactions:

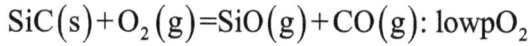

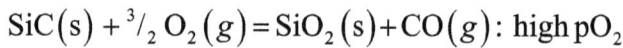

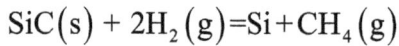

$$SiC(s) + O_2(g) = SiO(g) + CO(g): \text{low} pO_2$$

$$SiC(s) + {}^3/_2 O_2(g) = SiO_2(s) + CO(g): \text{high} pO_2$$

$$SiC(s) + 2H_2(g) = Si + CH_4(g)$$

$$SiC(s) + 2H_2O(g) = SiO(g) + CO(g) + 2H_2(g)$$

Silicon carbide is thermodynamically unstable in O_2, H_2, and H_2O environments under certain conditions. However, the kinetics of these reactions can be affected by the formation of a protective layer of SiO_2 at high pO_2 such that SiC is stable in many corrosive environments. The stability of the passive layer then becomes critical to the stability of SiC. The O_2 pressure for the SiC active–passive transition was determined by Gulbransen and Jansson as shown in figure. At temperatures of 800–1000 °C, this transition occurs at O_2 pressures of 10–8 atm. Therefore, the pertinent reactions then become those between SiO_2 and specific gaseous.

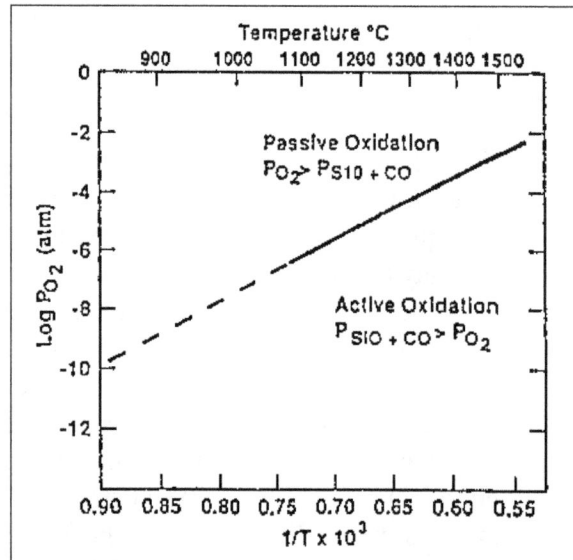

Transition pressures for SiC active–passive oxidation versus temperature, according to Gulbransen and Jansson.

or molten-salt environments, except in molten Li with a low O_2 activity, where SiO_2 is unstable. Some relevant reactions with SiO_2 are below:

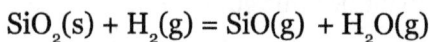

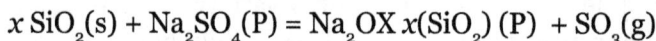

$$SiO_2(s) + H_2(g) = SiO(g) + H_2O(g)$$

$$x\,SiO_2(s) + Na_2SO_4(P) = Na_2OX\,x(SiO_2)(P) + SO_3(g)$$

where M is an alkali element such as Na and Li. A phase diagram for the Na_2O–SiO_2 system is shown in figure. Alkali elements such as Na and Li cause a breakdown of the passive SiO_2 film by the formation of low-melting alkali silicates such as those which occur at 800 °C in the Na_2–SiO_2 system and 1024 °C in the Li_2O–SiO_2 system. The eutectic temperature in the Li_2O–SiO_2 system is about 250 °C higher than that in the Na_2O–SiO_2 system and, therefore, Li is expected to have less effect on the passive film on SiC at temperatures below 1000 °C than Na.

A summary of the behavior of SiC in gas–molten-salt environments as presented by McKee and Chatterji is shown in figure. Passivation occurs at high pO_2, and active oxidation (formation of gaseous SiO) occurs at low pO_2. A basic salt or salt melt with a low pO_2 at the salt–SiC interface will cause active corrosion, as depicted by reaction scheme and. McKee and Chatterji suggest that SiC will not react with H_2; however, more recent analysis by Herbell et al. indicates that the equilibrium partial pressure of CH_4 is 10^{-4} atm at temperatures of 850–1400 °C for a H_2 pressure of 1 atm. While McKee and Chatterji measured the sample weight change in the test environment as a function of time in 1 atm of H_2 at 900 °C, they observed no reaction between SiC and H_2. However, their gas may have had sufficient O_2 or H_2O to cause passivation.

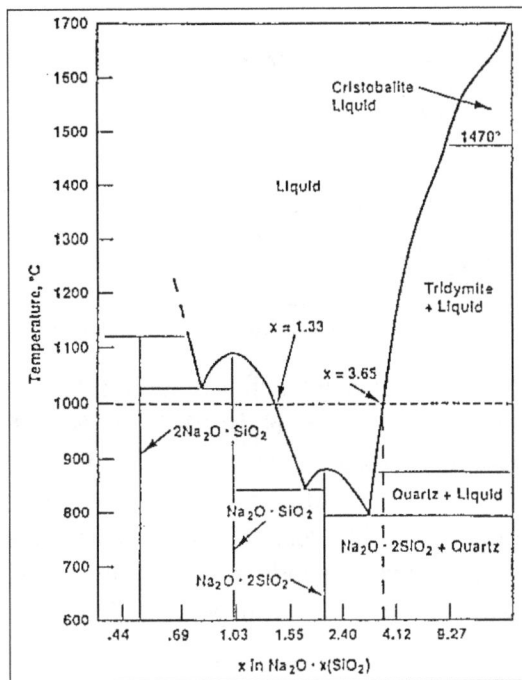

Phase diagram for the system Na_2O–SiO_2.

Herbell et al. calculated the SiO(g) partial pressure for Equation $SiO_2(s) + H_2(g) = SiO(g) + H_2O(g)$ with 1 atm of H_2 containing 1 ppm of H_2O to be 10^{-7} atm. Therefore, it would appear that a very small amount of H_2O mixed with O_2 to promote SiO_2 formation would be sufficient to cause a significant reduction in the reaction rate of SiC.

Jacobson evaluated the kinetics and mechanisms of the corrosion of sintered α-SiC in molten salts at 1000 °C. In the reaction of Na_2SO_4/O_2 with SiC, the reaction occurred primarily in the first few hours with the formation of a protective SiO_2 layer. This observation was demonstrated with results showing the total weight of the corrosion products/unit area reaching 6 mg/cm² (SiO_2 + Na_2O–$x(SiO_2)$) after a few hours and remaining constant up to 20 h. Jacobson and Smialek also noted that SiC is subject to pitting corrosion in molten salts. Pits occurred at structural discontinuities and bubbles formed during the formation of SiO_2. Pitting corrosion is detrimental because it demonstrated that the passive SiO_2 layer has been degraded and because the pits act as flaws resulting in reduced fracture strength.

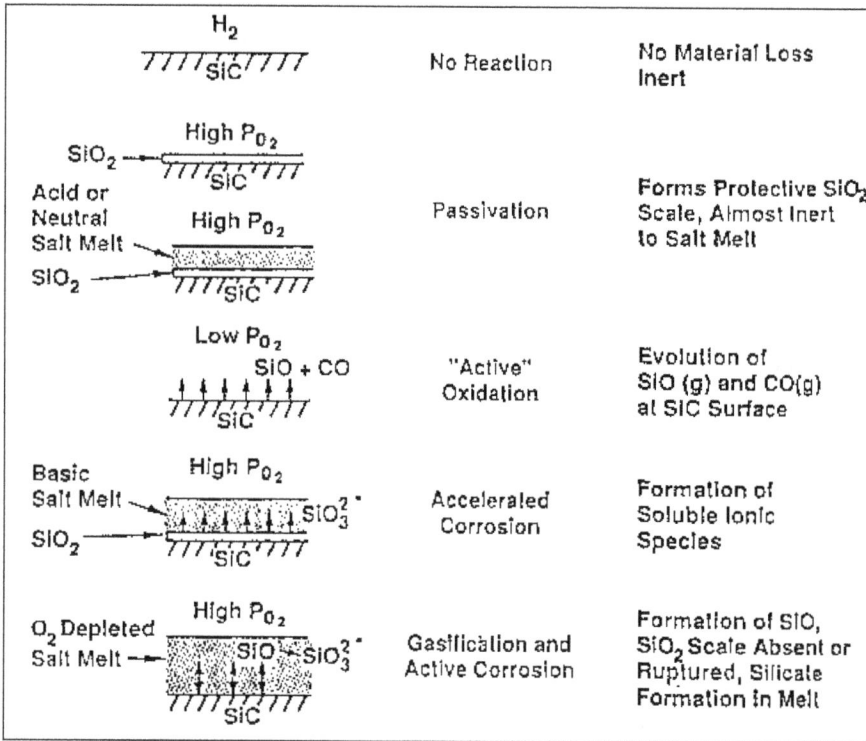

Possible modes of behavior of SiC in gas–molten-salt environments.

Corrosion measurements of SiC immersed in a crucible of molten Na_2SO_4 were conducted by Tressler et al. Weight change results for both SiC and Si_3N_4 in 100% Na_2SO_4 at 1000 °C are given in figure, where it is evident that considerable corrosion occurs for both materials, with SiC showing a higher corrosion rate than Si_3N_4. Both materials exhibited a weight loss because the reaction product was removed prior to measuring the weight change. An in situ measurement would probably exhibit a weight gain up to the time at which reaction product spallation occurs. Corrosion data obtained on SiC–Si_3N_4 and SiC–SiC composite materials by Henager and Jones is also presented in figure.

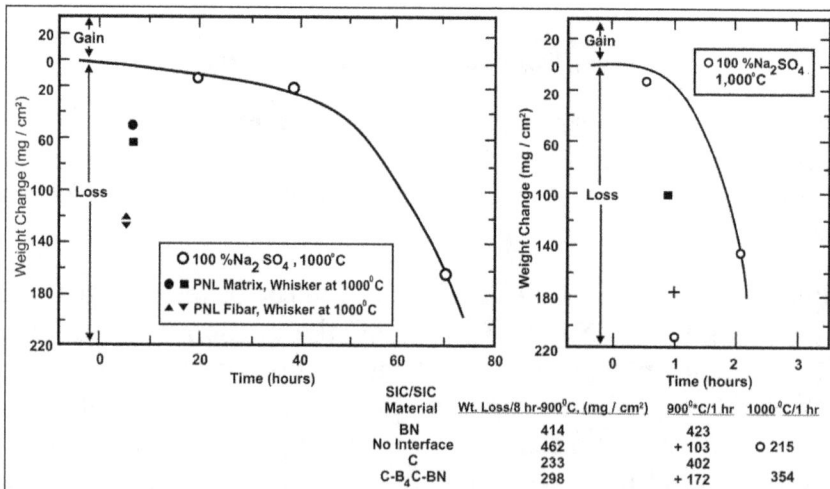

SIC/SIC Material	Wt. Loss/8 hr-900°C. (mg / cm²)	900°C/1 hr	1000 °C/1 hr
BN	414	423	
No Interface	462	+ 103	O 215
C	233	402	
C–B_4C-BN	298	+ 172	354

Weight change versus exposure time to Na_2SO_4 at 1000 °C for several ceramic composites.

The corrosion rate of the composite material was found to exceed that of the monolithic material for a Nicalon–SiC-reinforced, hot-pressed Si_3N_4 and a Nicalon– SiC reinforced CVI SiC. A SiC-whisker-reinforced, hot-pressed Si_3N_4 exhibited a similar corrosion rate to the unreinforced hot-pressed Si_3N_4. The activation energy for the corrosion of the fiber-reinforced material was 880 kJ/mol, whereas the whisker-reinforced matrix material had an activation energy of 1280 kJ/mol. The difference in corrosion rate and activation energy for the fiber-reinforced material was associated with preferential corrosion of the Nicalon fiber. The fiberreinforced material had a carbon interfacial layer between the fiber and matrix, and the whisker-reinforced material had a thin amorphous glass layer. However, the difference in corrosion behavior is not thought to be associated with the interfacial layer but with the structure of the Nicalon fiber following hot pressing. As produced, Nicalon has an amorphous/microstryalline structure while the hotpressed structure was crystallized with evidence for a Mg silicate in the grain boundaries. The source of the Mg was the MgO sintering aid used for the hotpressing process. Corrosion of the Nicalon fiber along the glass-enriched grain boundaries probably accounted for the high corrosion rate of the Nicalon-fiberreinforced composite material.

The corrosion rate of a SiC–SiC composite produced by the CVI process was also greater than that of monolithic SiC. Microstructures of CVI SiC–SiC composites made by the ORNL process and the Du Pont process both exhibit porosity, which exists in materials made by this process. Composites containing fibers coated with C, BN, C–B_4C–BN, and no coating were evaluated. The corrosion rate was highest for the material made with fibers with no coating and the BN coating, whereas composites made with C and C–B_4C–BN coatings exhibited smaller corrosion rates. However, the corrosion rate of these materials at 900 °C was considerably greater than that of monolithic material at 1000 °C. Matrix corrosion contributed the majority of the weight change, and while some fiber corrosion was noted, although this was a relatively small factor compared to the matrix corrosion. The open porosity and penetration of the molten salt into the composite are considered the primary factors causing the high corrosion rates for the SiC–SiC composite.

Fox et al. found similar results in burner rig corrosion studies of CMCs to those obtained in molten salt. The burner rig tests utilize a hot-combustion gas and would be expected to differ from a molten Li or He gas environments; however, molten deposits are one mechanism for corrosion in hot-combustion gas studies. Fox et al. evaluated Si_3N_4 reinforced with 30 vol% SiC whiskers and fibers. The method of procesing for the whisker-reinforced materials was not reported; however, it was probably hot pressed, whereas the fiber-reinforced material was reaction bonded and had a residual porosity of 30%. Tests were conducted for 40 h at 1000 °C in an environment containing 2 ppm Na added as NaCl. Monolithic Si_3N_4 exhibited a 10% decrease in the fracture stress and the whisker-reinforced material a 35% decrease. The fiber-reinforced material exhibited an excessive corrosion rate, but no fracture studies were conducted. The high residual porosity in the reaction-bonded material and the free C from the carboncoated fibers were postulated as the cause for the high corrosion rate in the fiberreinforced material. In the molten-salt corrosion studies, the more porous CVI composite material also exhibited a higher corrosion rate than did the hot-pressed Si_3N_4/fiber-or whisker-reinforced-material.

Effects of corrosion on the mechanical integrity of SiC–SiC structural components is an important factor in the durability of components constructed of SiC/SiC. Decreases in the fracture strength would be possible by corrosion penetration along grain boundaries. Smolten- salt-induced pits were responsible for up to 50% of the observed strength reduction of SiC after 48 h at 1000 °C in Na_2SO_4–SO_3. A relationship of fracture stress versus (pit depth)—2 illustrated the flaw-induced fracture relationship. Henshall et al. demonstrated that combustion gases containing alkali salts also contribute to accelerated subcritical crack growth of monolithic SiC. The crack velocity in the hot-combustion gas was several orders of magnitude faster at a 50 K lower temperature than in air. The primary mechanism for this accelerated crack velocity is penetration of alkali ions into the grain-boundary glass phase and the reduction in the viscosity of this phase. The reduced viscosity causes increased crack tip creep rates and creep damage.

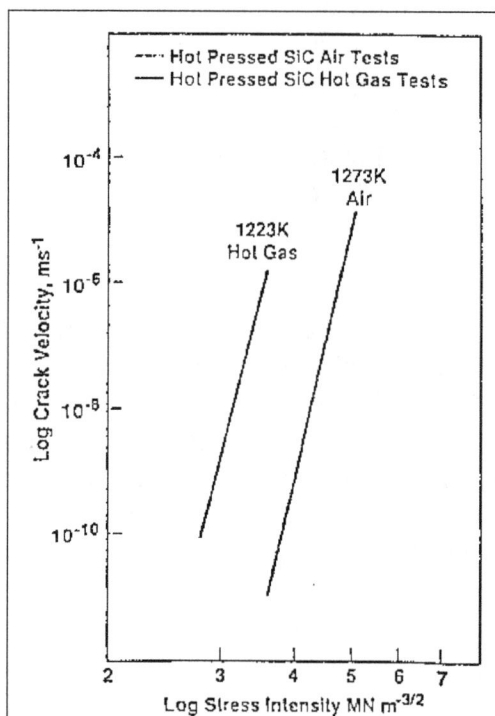

Crack velocity versus stress intensity for monolithic SiC exposed to air and hot gas environments.

Stability of SiC in Mildly Oxidizing Environments

Spear et al. measured the oxidation rate of hexagonal α-SiC platelets with both (0001) C and (0001) Si faces. The only solid corrosion product observed was SiO_2 [Eq. $(SiC(s) + {}^3/_2 O_2(g) = SiO_2(s) + CO(g)$: high pO_2)], with CO(g) formed at the SiC–SiO_2 interface. The O_2 pressure was 1023 to 1 atm at 1200–1500 °C, which is above the active–passive transition; thus, the presence of SiO(g) would not be expected. An activation energy of 120 kJ/mol was found for SiC, which is very comparable to the value of 112 kJ/mol found for Si. Silicon nitride has a much higher activation energy of 486 kJ/mol because of the formation of a silicon oxynitride phase between the Si_3N_4 and the SiO_2 phases. A growth rate of 1.1 nm/min was determined for SiC at 1300 °C in 1 atm of O_2. Oxygen diffusion was postulated as the rate limiting process up to 1350 °C, above which ionic oxygen diffusion was rate controlling.

The protective properties of the passive SiO_2 film are very sensitive to the impurities of the SiC from which the film is formed. Comparison between the oxidation behavior of several batches of SiC to that of very high-purity SiC produced by chemical vapor deposition was made by Fergus and

Worrell. They observed an activation energy of 142 kJ/mol for CVD SiC and 217 kJ/mol for sintered α-SiC. This compares with the value of 120 kJ/mol reported by Spear et al. for single-crystal SiC platelets. Both the CVD and SiC platelets are of much higher purity than the sintered α-SiC to which sintering aids were added. Vaughn and Maahs in a review of the active–passive transition for SiC showed that this transition was quite variable depending on the source of the SiC and other experimental conditions such as the gas flow velocity. The Gulbransen data was very close to that determined theoretically, but most other materials exhibited much higher transition temperatures, presumably due in part to variable impurity concentrations in the materials.

Results of oxidation studies of ceramic composites have not been reported, although Luthra has presented some theoretical concepts on the oxidation of ceramic composites. Luthra assessed the gas-phase diffusion of oxygen through microcracks, solid-state diffusion of oxygen through protective oxide layers, formation of gaseous reaction products, and the combined reaction of the matrix reinforcement and oxygen. In particular, the diffusion of oxygen through microcracks to react with the reinforcing fiber or the interface between the fiber and matrix is a definite possibility for SiC–SiC composites produced by the CVI process because they have at least 10% residual porosity. Thermal cycling is also likely to induce matrix microcracking because of the thermal expansion mismatch between the reinforcing fiber and the matrix. For SiC fibers in a CVI SiC matrix, this mismatch will be small, but it is not zero. Luthra's analysis is primarily for oxide matrix composites, where the oxygen will not react with the matrix, whereas for SiC where the oxygen could react with the matrix, the transport would be considerably slower.

The structural behavior of CMCs in an inert gas environment will be dependent on (a) whether the material is undergoing active or passive oxidation, (b) the stability of the fiber–matrix interfacial layer in O_2, and (c) the stability of the fiber in O_2. Kim and Moorhead have recently evaluated the flexural strength of α-SiC at room temperature following a 10-h exposure to Ar–O_2 at 1400 °C. Flexural strengths following exposure to Ar–O_2 mixtures with O_2 partial pressures of 7×10^{-7} to 2×10^{-4} MPa showed a strength reduction with increasing pO_2 up to about 2×10^{-5} MPa and a complete restoration of the as-polished strength at p$O_2 > 2 \times 10^{-5}$ MPa. The minimum strength (60% of unexposed material) at 2×10^{-5} MPa was the result of active corrosion creating strength-reducing flaws along grain boundaries and at pits. Easler et al. found that the room-temperature flexure strength both increased and decreased upon exposure to air at 1370 °C with and without an applied load. The strength increases were thought to result from oxidative blunting, whereas the decreases which occurred under load were due to subcritical crack growth producing larger flaws. Minford et al. also found that stress enhanced the uptake of O_2 into SiC. Oxygen penetration occurred along cation enriched grain boundaries. A stress intensity threshold of about 1 MPa. m$^{1/2}$, below which crack blunting occurred and above which crack extension occurred was found.

Oxidation can alter the dynamic crack growth behavior of SiC as well as reduce the fracture strength due to corrosion-induced flaws. A reduction in the K_{th} of α-SiC after being loaded to different stress intensities for 4 h at 1200 °C and 1400°C was reported by Minford et al. They reported a reduction in the K_{th} from 2.25 in the unoxidized to 1.75 MPa. m$^{1/2}$ in the oxidized conditions at 1200 °C and from 1.75 to 1.25 MPa . m$^{1/2}$ at 1400 °C for samples tested in air or preoxidized. They concluded that the O_2 caused a change from diffusion-controlled crack growth to viscous cavity growth and linkage with a corresponding reduction in K_{th}. The K_{th} was defined as the transition between crack blunting and growth as determined from the fracture strength following the 4-h static loading at various K values. McHenry and Tressler found that the K_{th} and subcritical crack growth rate was independent of the pO_2 for pressures ranging from 10—8 to 10^{-4} atm and temperatures of 900–1100 °C. The crack growth rate exhibited an Arrhenius temperature dependence with an activation energy of 20 kcal/mol, which is consistent with viscous flow of a grain-boundary glass phase. The lack of a dependence on pO_2 does not necessarily indicate that O_2 did not induce crack growth but that the effect of O_2 was saturated at pressures above 10^{-8} atm.

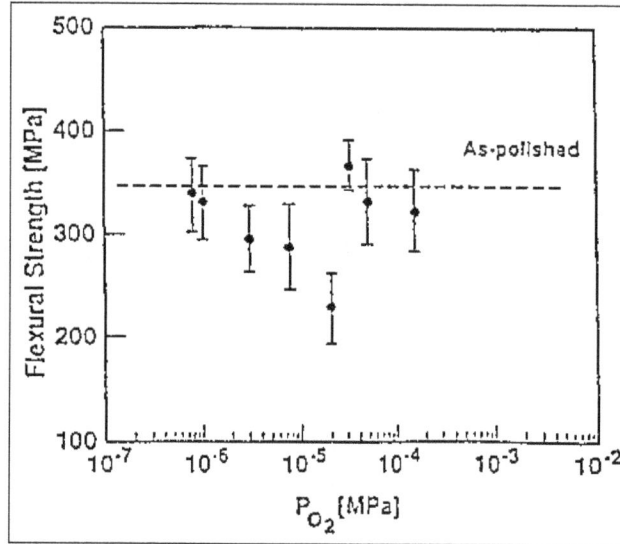

Room-temperature flexural strength of sintered α-SiC after exposure for 10 h at 1400°C in Ar atmospheres with various pO_2.

Oxidizing Environments

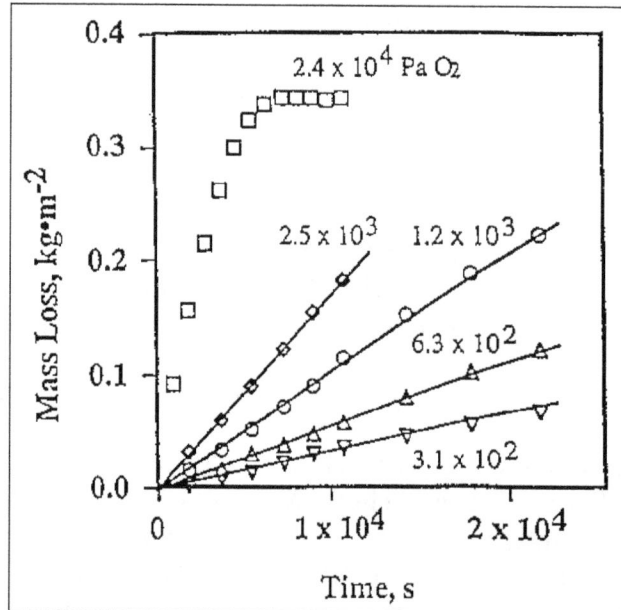

Mass loss versus exposure time for SiC–SiC with a C interphase exposed to various O_2 partial pressures at 1100 °C.

A key feature of the oxidation behavior of SiC–SiC in O_2 at pressures of 2 × 104 Pa (atmospheric pressure of O_2) observed by Windisch et al. is that only a weight loss was observed. No SiO_2

formation occurred in any of the materials with graphite interphases, although some boron-containing glass phase was observed for the material with a BN interphase. The kinetics of mass loss is shown as a function of pressure and temperature in figures and Complete burnout of the graphite interphase occurred in less than 10^4 s in the small test samples at a pressure of 2.4×10^4 Pa and a temperature of 1373 K. It is also clear that the reaction rate increases with increasing temperature. An activation energy of about 50kJ/mol which could be explained as diffusion controlled through a boundary layer or as being reaction rate controlled.

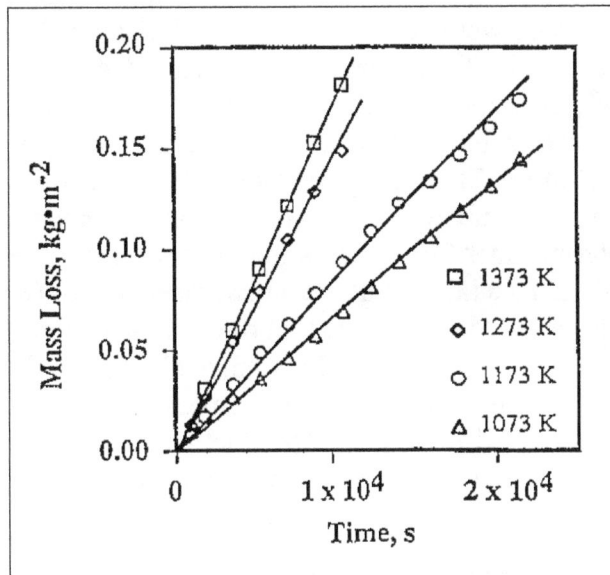

Mass loss versus time for SiC–SiC with a C interphase exposed at 2.5×10^3 Pa O_2 at various temperatures.

An interphase recession rate was determined from the weight-loss measurements and by direct physical measurement. Both methods gave very similar recession rate equations with the physically measured equation as follows:

$$\log (RR) = 0.9 \log(pO_2) - 9.9$$

It should also be noted that only weight loss was observed over the temperature range 1073–1373 K, which borders on the temperature range (873–1073 K) suggested by Evans et al. for the pest phenomena.

Oxidation of SiC–SiC composites with carbon interphases can also result in the formation of SiO_2 and a weight gain following an initial weight loss or a reduced weight loss with increasing temperature. Tortorelli et al. observed an initial weight loss followed by a weight gain in a Nicalon-fiber reinforced SiC–SiC composite with a 0.3-μm-thick graphite layer exposed to dry air (pO_2 of 2×10^4 Pa) at 1223 K. Following the initial weight loss from oxidation of the carbon, they found that SiO_2 formation occurred within the interfacial region previously occupied by the carbon. For the Nicalon-reinforced composite material, complete carbon depletion occurred within 15 min at 1223 K, followed by weight gain from SiO_2 formation. Unal et al. observed their largest weight loss (5%) at 1223 K for an exposure of 50 h in dry O_2 and a decreasing weight loss with increasing temperature up to 1673 K (2%). Their material was a Nicalon-reinforced SiC–SiC with a 0.5-m-thick fiber–matrix carbon interphase. Kleykamp et al. observed the following reactions of air with SiC-fiber-reinforced SiC composites: (a) oxidation of free carbon at temperatures of 800–965 K and (b) a fast exothermic reaction and weight gain beginning at 1073 K and containing up to 1773 K and up to times of 1 h followed by (c) the diffusion-controlled oxidation of bulk SiC. Sebire-Lhermitte et al. identified the presence and location of SiO_2 formation in SiC–SiC composites using transmission electron microscopy. They noted the presence of 15-nm-thick SiO_2 layers at both the fiber carbon and matrix–carbon interfaces following an exposure of 1 h at 1123 K in air.

That Windisch et al. only observed a weight loss while others observed a weight gain following the weight loss could be the result of lower O_2 pressures, exposure time, and perhaps, carbon-layer thickness. The lower O_2 pressure would reduce the SiO_2 formation rate and therefore the chance for a measurable weight gain during the 5-h exposures. Tortorelli et al. used exposures of up to 150 h, whereas Unal et al. used 50-h exposures. Measurable SiO_2 formation at $pO_2 < 2 \times 10^3$ Pa would need a much greater exposure time than that used by Windisch et al. and even greater than the time used by Tortorelli and More. The existence of subcritical crack growth, as described in the next section, that coincides with only weight loss or interphase removal without the embrittling effect of SiO_2 or other solid-product formation is the primary difference between an interphase removal mechanism (IRM) and oxidation embrittlement mechanism (OEM) of crack growth. The oxidation results of Windisch et al. demonstrate that the results of Henager et al. at temperatures ranging from 1073 to 1473 K and $pO_2 < 2 \times 10^3$ Pa occurred by IRM only. The OEM is dependent on SiO_2 formation, which depends on O_2 pressure, temperature, and time. The interfacial layer thickness may also impact this regime if a solid product can seal off the interface from further reaction. A clear demonstration of this possibility has not been presented but remains as an open issue needing further evaluation.

The effect of oxygen on the subcritical crack growth velocity of SiC–SiC is clearly demonstrated by the data given in figure. Oxygen has little effect on the midpoint displacement (i.e., crack velocity) for about 2×10^4 s, but a marked increase in the crack velocity is noted for longer times. These tests were performed in the O_2 pressure, temperature, and time regime where only weight loss was observed during oxidation studies. Therefore, the embrittling effect of a solid reaction product should not be a factor, but only the effect of fiber creep and interfacial removal should have contributed to the crack growth rate. However, even if SiO_2 or other glassy phases were present, they would have low viscosity at very high temperatures and would not likely affect the crack growth behavior or cause brittle fracture.

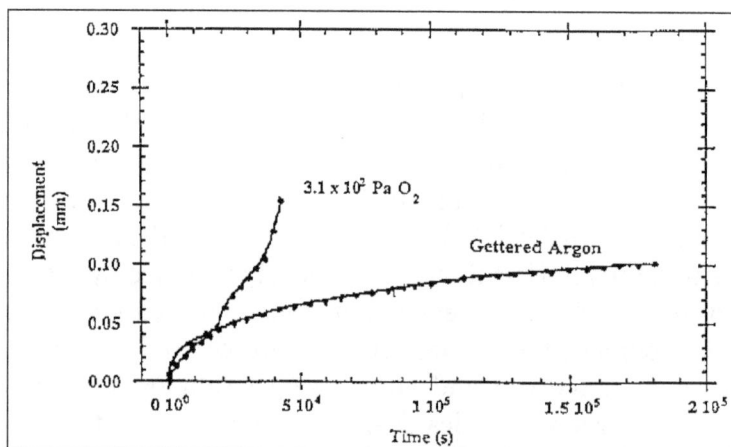

Midpoint displacement of a single-edge notch-beam specimen of SiC–SiC with a C interphase in gettered argon and 3.1×10^2 Pa of O_2 + Ar at 1200 °C.

The dependence of the crack velocity on O_2 partial pressure up to 10% that of air is given in figure for tests at 1373 K. There is a sharp increase in the crack velocity at low pressure and a slower increase from pressures of about 0.25×10^2 Pa up to 2.5×10^3 Pa. Material with a BN interface exhibited about a factor-of-10 slower crack velocity. Some glass-phase formation was noted in this material as a result of these exposures, but there was no evidence that the crack growth behavior was affected by the presence of this glass phase.

The IRM has been observed to cause crack growth in SiC–SiC at temperatures of 1073–1473 K at pressures ranging from 2×10^2 to 2×10^3 Pa for stressed samples. Weight-loss measurements suggest that the IRM operates over these same temperatures and pressures and at 1373 K and a pressure of 2×10^4 Pa. An example of the crack velocity versus crack length for tests on specimens with Hi-Nicalon-reinforced material tested in gettered Ar and Ar + 2×10^2 Pa of O_2 is given in figure. The acceleration in the crack velocity induced by the presence of O_2 is clearly shown by these data, whereas the effect of temperature on the Ar + O_2 test is only apparent after a crack extension of 3.5 mm.

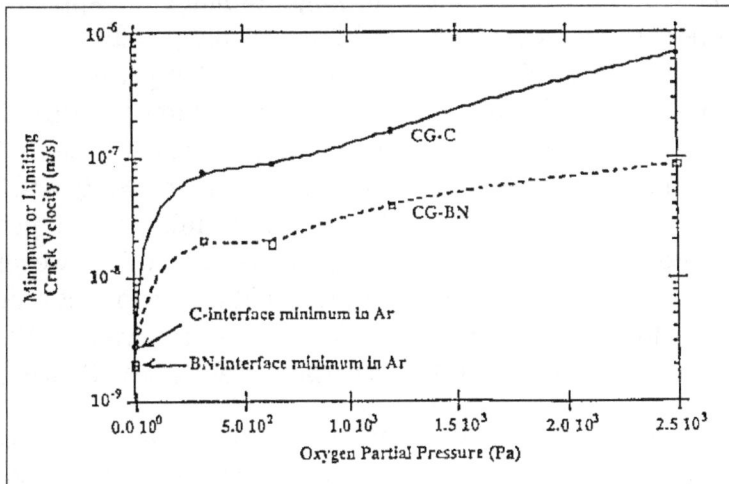

Minimum crack velocity versus O_2 partial pressure for SiC–SiC with a C or BN interphase exposed at 1100 °C.

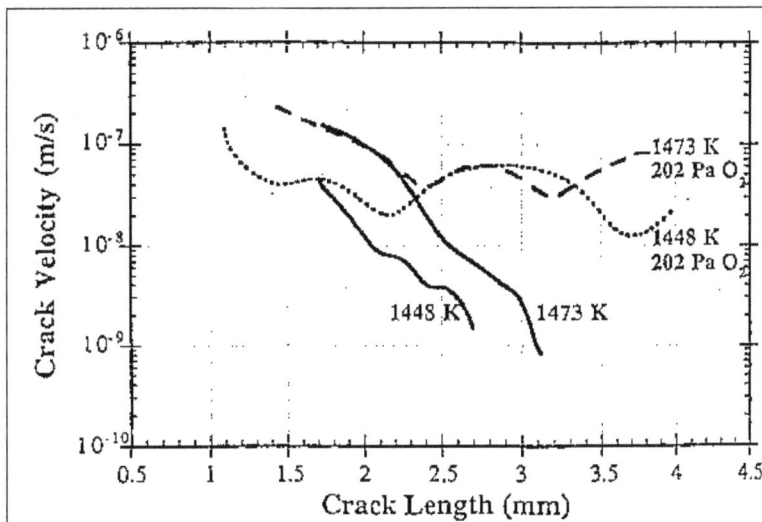

Crack velocity versus crack length for SiC–SiC with a C interphase exposed to 202 Pa of O^2 1 Ar (dashed curves) or gettered Ar (solid curves) at 1175 °C (1448 K) and 1200 °C (1473 K).

Dynamic (sample stressed during exposure) OEM has been observed to occur in SiC–SiC at temperatures up to 1073 K, and 1223 K in air (O_2 pressure of 2×10^4), whereas static (sample unstressed during exposure) OEM was also observed in room-temperature tests following elevated-temperature exposures at 1223 K in air (23) and up to 1673 K in dry O_2 at a pressure of 105 Pa. Heredia et al. reported an upper pest temperature of 1073 K for SiC–SiC tested at elevated temperature in air, whereas Lin and Becher and Raghuraman et al. observed OEM in air at 1223 K. Tortorelli et al. and Unal et al. found the OEM to operate in room-temperature tests following elevated-temperature exposure tests conducted in air without the application of stress. The upper temperature limit for the dynamic operation of OEM is between 1073 and 1223 K, and the formation of a glass phase at temperatures greater than 1223 K can cause OEM to occur when specimens are tested at lower temperatures. Because the OEM results from the formation of a brittle glass phase, this mechanism must depend on the growth rate and viscosity of the glass phase. The growth rate will increase with increasing temperature and pO_2, but the viscosity decreases with increasing temperature. Therefore, there must be a temperature at which the effectiveness of OEM is maximum.

The results of Heredia et al, Lin and Becher, and Raghuraman et al. appear to define the upper temperature and O_2 pressure limits for OEM in SiC–SiC at 1073–1223 K and O_2 pressures of 2×10^4 Pa and above. However, the IRM appears to operate over temperatures of at least 1073–1473 K at O_2 pressures of 2×10^3 Pa and below. It may also operate at temperatures below 1073 K, within the OEM range, at low pressures, but this has not been observed because the crack velocities are too low for experimental measurements. Bouche tou et al. and Frety and Boussuge observed a degradation of the mechanical properties of SiC–SiC containing cracks produced by stress or thermal gradients and exposed to an oxidizing atmosphere at 773 K. however, OEM is expected at temperatures below about 1223 K, although 773 K would be the lowest reported occurrence to OEM. The low activation for carbon oxidation (50 kJ/mol), as reported by Windisch et al, would result in a small decrease in the oxidation rate of a carbon interphase with decreasing temperature.

The transition from OEM to IRM is not only a function of temperature and oxygen pressure but also of the thickness of the carbon interphase layer between the fibers and matrix. Filipuzzi et al. and Cawley noted that interphases with a thin carbon layer (e.g., 0.1 μm) were quickly sealed by SiO_2, whereas the carbon was totally oxidized before glass formation in material with a thicker interphase region (e.g. 1 μm). The interphase thickness would not alter the temperature dependence of the OEM to IRM transition but would lower the pO_2 and shorten the time for OEM to occur in favor of IRM. A schematic of the matrix, interphase, and fiber oxidation process shown in figure from Cawley shows the competition between interphase oxidation and SiO_2 formation on the matrix and fiber that leads to sealing off of the interphase region. Huger et al. measured the oxidation rates for Nicalon NLM202 fibers exposed to air at temperatures ranging from 700 °C to 1200 °C. After 100 h, the weight change was about five times greater at 1200 °C relative to 700 °C with a 1-μ m-thick glass layer forming after 100 h at 1000 °C. Clearly, the oxide layer can grow sufficiently thick to seal the interphase region and to perhaps to act as a crack initiator which could reduce the fiber strength.

The environmental stability of composites with silicon-based matrices, such as blackglass, nitrides, and carbides have also been evaluated by several authors. In studies of material with Nextel 312 fibers in a matrix of Allied Signal Blackglass™ with a BN interphase, Campbell et al. measured the weight change and bend strength of material exposed to dry air, air + 3% H_2O, dry air + 80 ppm KCl vapor, and air containing 3% H_2O + 80 ppm KCl vapor. The exposures were conducted for variable lengths of time at 700 °C and 5 h at 900 °C. The composites show only continuous weight loss with time for exposures to dry air at 700 °C, but a weight gain following a small initial weight loss in KCl containing environments. The weight gain was identified as resulting from the formation of alkali silicates. Vaidyanathan et al. also studied the effects of oxidation on the mechanical properties of the Nextel 312/BN/blackglass composite but for times of 20–1000 h at 600 °C. They observed a 50% reduction in strength after 500 h at 600 °C and concluded that times greater than 200 h under these conditions were a concern for the durability of the composite. Oxidation resulted in increased fiber pullout, consistent with IRM, but at a much lower temperature than IRM in a SiC–C–SiC composite. The degradation may have resulted from the weakening of the fiber–matrix interface.

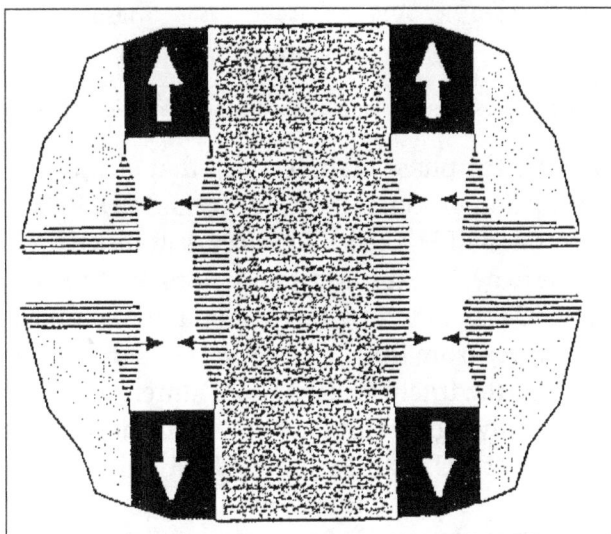

Schematic of O$_2$ reaction with the C interphase and interphase recession and with the SiC matrix and fiber to form SiO$_2$.

The instability of the fiber–matrix interphase in SiC–SiC composites has led to the evaluation of coatings for these materials. Fox evaluated the oxidation resistance of three coatings on SiC–SiC composites: (a) CVD SiC, (b) a particulate-based sealant with a CVD SiC outer layer, and (c) a boron-rich inner layer and CVD SiC outer layer. Oxidation studies were conducted in dry oxygen at 981–1316 °C. All three provided protection for up to 100 h, and the CVD SiC was the most protective. Oxygen diffusion through the SiO$_2$ that formed on CVD SiC was considerably slower than through the glass layer that formed on the other two coatings which contained B. Lee and Miller evaluated the stability of mullite coatings with a refractory oxide barrier coating on SiC–SiC exposed to air at 1200–1400 °C. The sample temperatures were cycled every 1, 2, or 20 h from 1200 °C and 1300 °C to room temperature and every 1 h from 1400 °C to room temperature. The mullite–refractory oxide composite coating exhibited improved adherence and oxidation resistance relative to a straight mullite coating.

Hydrogen-containing Environments

Herbell et al. have evaluated the thermodynamic stability of SiC in pure H$_2$ at 1 atm, as shown in figure. The primary gaseous reaction produce is CH$_4$ as described by Equation $(SiC(s)+2H_2(g)=Si+CH_4(g))$, whereas other reactions which produce SiH$_4$ and SiH are also possible at temperatures as low as 900 °C.

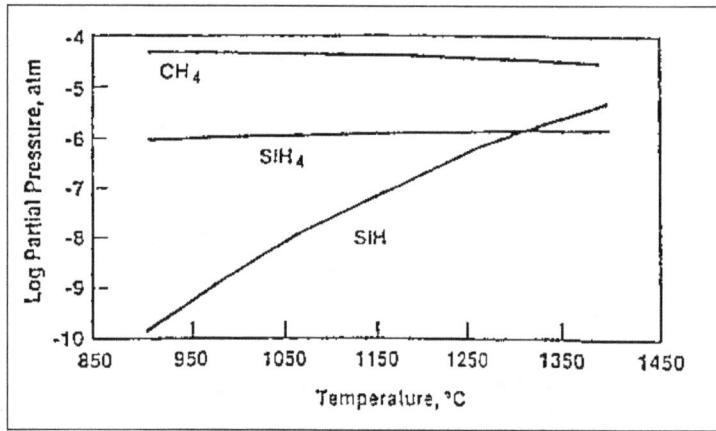

Gases for equilibrium partial pressures of reaction products for reaction of SiC with pure H_2 at 1 atm.

A small amount of H_2O can alter the phase stability such that at 1400 °C and about 1000 ppm of H_2O, the dominant gaseous reaction products become SiO and CO. Results for lower temperatures were not reported, but the reaction of H_2O with SiC occurs at much lower temperatures, so similar reaction products would be expected at lower temperatures. No loss in the room-temperature flexural strength of sintered SiC was noted by Herbell et al. for samples exposed to H_2 saturated with H_2O for 100 h at temperatures from 800 °C to 1400 °C. In dry H_2 (25 ppm H_2O), Hallum and Herbell noted a 33% decrease in the fracture strength of sintered SiC after exposures of 500 h at 1100 °C and 1300 °C. A statistically significant decrease in flexural strength was also observed after 50 h at 1000 °C. The stability of SiC in an Ar–H_2O–5% H_2 mixture was calculated by Jacobson et al. in the same manner as the H_2–H_2O mixtures, with the result shown in figure. Except for the lower gas pressures and the shift in the relative activities of SiO and CH_4 in region III, the results are essentially identical. At 1300 °C, Jacobson et al. measured a weight loss of around 1 mg/cm² after 24 h in a region II Ar–H_2–H_2O mixture.

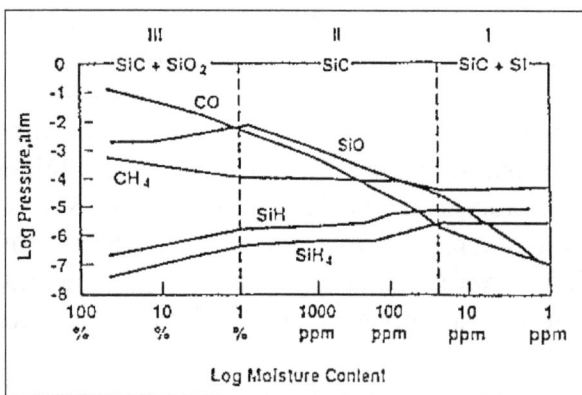

1400 °C stability of SiC in H_2 + H_2O at 1 atm.

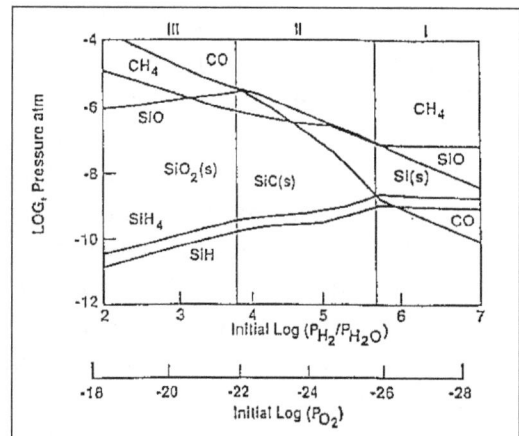

Thermodynamic analysis of SiC + 5% H_2–Ar at 1300 °C. All pressures are in atmospheres.

Hydrogen may also react with the carbon interphase to form CH compounds. This reaction would be in addition to the direct reaction with matrix and fiber as described by Herbel et al. Springer et al. evaluated the reaction of Ar + H_2 environments on the weight change of SiC–SiC composites which had a carbon fiber–matrix interphase. They used a thermogravimetric analyzer to study the weight change at 1000–1200 °C with Ar + 0.1% H_2 and Ar + 1.0% H_2. They found that the reactivity of the carbon interphase to H_2 was substantially less than to O_2. For instance, Ar + 100 ppm O_2 produced a weight loss 26 times greater at 1000 °C relative to Ar + 0.1% H_2. Nightingale found an activation energy of 65 kcal/mol for H_2 reacting with bulk graphite, whereas Springer et al. found activation energies of 18 and 34 kcal/mol for 0.1% and 1.0% H_2, respectively. The carbon interphase material is a mixture of amorphous carbon and graphite so that the lower activation observed for the carbon interphase material could be the result of the lower stability of the interphase material relative to bulk graphite.

Oxide Matrix Composites

The chemical instability of the SiC in the presence of alkali elements and the fiber–matrix interphase in SiC–SiC composites in oxidizing environments are factors that encourage the development of oxide matrix composites. Much of the high-temperature corrosion data for oxide matrix composites exists for particulate-, whisker-, or platelet-reinforced material, which have not been optimized for strength and toughness. Oxide fiber development has progressed to the state where continuous-fiber composites are being produced; however, there is no high-temperature oxidation data for these materials. Examples are alumina and alumina–YAG matrix composites reinforced with Nextel 610 and 720, as reported by Goettler. Interphase layers of $ErTaO_4$ or $CaWO_4$ are being evaluated for producing fiber pull out and fracture resistance. However, the stability of these interphase materials in oxidizing, reducing, or salt environments has not been evaluated. Even though the matrix and fiber may exhibit excellent behavior in oxidizing environments, uncertainty about the composite chemical stability remains. Also, the high-temperature strength of oxide fibers is less than SiC fibers, such that further improvements in strength must occur before continuous-fiber oxide–oxide composites are attractive for high-temperature applications.

Borom et al. have examined the oxidation behavior of Al_2O_3 reinforced with SiC and $MoSi_2$ particles and SiC whiskers. The particulate volume fractions ranged from 10% to 30% and tests were conducted at 1200–1500 °C in air; the oxidation rate was determined by weight change and reaction layer thickness. Both SiC and $MoSi_2$ form protective SiO_2 layers when oxidized as bulk materials. Borom et al. reported a 15-fold increase in the oxidation rate of these phases when incorporated into an Al_2O_3 matrix. This increase was postulated as resulting from the volume change of the reaction product that forms on the composite and the thermal expansion mismatch of the reaction product with the composite. Both of these factors were less favorable for the composite as compared to bulk SiC and $MoSi_2$. Larger volume fractions of these phases produced a large volume fraction of mullite in the reaction scale and this was favorable because the silica in the mullite will produce a more viscous scale that will allow more stress relaxation and accommodation for mismatch stresses a mullite matrix is preferred because the reaction product will contain aluminosilicate plus mullite, which will flow and relax thermal mismatch stresses.

The bend strength of SiC-whisker-reinforced (28 vol%) Al_2O_3 was found by Leaskey et al. to increase by 33% when oxidized in air at 1600 °C for 15 min. Composites with SiC particle reinforcement showed a 66% improvement in the bend strength following an oxidizing treatment of 2 h at 1600 °C.

The improved properties are the result of the oxidation of the SiC reinforcement to produce a compressive surface layer. The following conditions were necessary for this improvement: (a) a sufficiently large SiC content to produce a continuous oxide surface layer, (b) oxidation conditions that produce a low porosity layer with a critical thickness, and (c) elimination of large flaws in the bulk of the material.

The tensile strength of Nicalon-fiber-reinforced Al_2O_3 following heat treatment in air at 750 °C has been reported by Heredia et al. This material contained 0°/90° fiber orientation with a BN–SiC interphase. The room-temperature tensile strength was found to decrease from about 250 MPa to about 120 MPa following exposure to air at 750 °C for 24 h. The formation of a glass phase on the Nicalon fiber was suggested as the cause of the observed oxidation embrittlement.

Corrosion studies of SiC-reinforced Al_2O_3 have been conducted in coal combustion environments by Watne et al. and Breder et al. In the study reported by Watne et al, the material produced by the Lanxide Corp. contained 50% SiC with 10% residual Si. Following a 100-h pilot-scale combustion test at about 1350 °C in a radiant zone of the furnace, the composite, in the form of a tube, was intact but had a 0.85-mm reduction in the wall thickness. This loss was suggested as resulting from erosion from slag flow on the tube. The original wall thickness was 5.25 mm. A smaller amount of loss was found for a tube placed in the convective pass region of the combustor where the temperature was about 1200 °C. Breder et al. exposed a similar tube made by Lanxide Corp. to coal slag obtained from two coal combustion plants. Exposures were conducted in a box furnace with the tube and coal slag at temperatures of 1090 °C, 1260 °C, and 1430 °C. Fracture tests were conducted on samples removed from the tubes following a 500-h exposure. The tube strengths were reduced by 20–45% at 1260°C depending on the type of slag.

Although Al_2O_3 is the most commonly used matrix for oxide matrix composites, composites with other oxides such as MgO, ZrO_2, and mullite have also been evaluated. The oxidation kinetics of SiC particulate-reinforced MgO has been examined by Hallum and Camey and Readey. Hallum studied MgO reinforced with 5, 10, and 15 vol% SiC particles or whiskers over the temperature range of 1100–1500 °C. The reaction-product thickness increased with the square root of time and was a function of the volume fraction of SiC in the composite. Mg cation diffusion was proposed as controlling the growth rate with a reaction layer formed by Mg cation diffusion through the reaction layer to the atmosphere where oxidation produced a columnar growth region. Camey and Readey identified three oxidation-product layers unlike the single layer observed by Hallum; however, they agreed with Hallum regarding the growth rate being controlled by Mg cation diffusion through the product layer. Luthra and Park evaluated the oxidation of SiC in mullite and alumina matrices and found parabolic rate constants that were three orders of magnitude larger than SiC. Xu et al. measured the effects of adding ZrO_2 to mullite on the oxidation of mullite–zironia–SiC composites. They found that the addition of ZrO_2 to mullite–SiC composites increased the reaction rate with oxygen. They rationalized this as being due to the increased diffusion rate of oxygen in the zirconia phase. A rapid "mode II" type of oxidation, where oxygen can penetrate deep into the sample before the outer region is completely oxidized, occurred at 1200–1400 °C and with the volume percent of ZrO_2 greater than 20%.

Glass Matrix Composites

Glass and glass–ceramic matrix composites are the most developed class of ceramic matrix composites. These composites are easier to prepare than SiC–SiC or oxide matrix composites and so have received further development and evaluation than other CMCs. The glass matrices employed in these composites include calcium–aluminosilicate (CAS), lithium–aluminosilicate (LAS), magnesium– aluminosilicate (MAS), and barium–magnesium–aluminosilicate (BMAS). There have been a number of microstructural, mechanical property and environmental effects studies of materials reinforced with Nicalon-type fibers.

High-temperature Air Environments

Alteration of the fiber–matrix interface is one of the primary effects of oxidation on glass matrix–Nicalon composites. Daniel et al. evaluated the oxidation of CAS–Nicalon composites over the temperature range of 375–600 °C in air for 100 h. They evaluated the change in the fiber–matrix interfacial properties with a nanoindentation push-down test and four-point bend tests. At exposure temperatures of 450 °C and above, the composites exhibited brittle failure with minimal fiber pullout. The transition from tough behavior with fiber pullout for lower temperature exposures to brittle fracture was associated with an increase in the fiber–matrix frictional shear stress. This increase in the frictional shear stress is accompanied by the loss of the fiber–matrix interfacial carbon layer and the resulting residual stress causing the matrix to apply a compressive stress to the fiber. This clamping stress on the fiber reduces fiber pullout and causes brittle type behaviour.

Microstructural evaluation of the fiber–matrix interfacial region of CAS and LAS–Nicalon fibers exposed to air at 600 °C or 900 °C have been reported by Cooper and Chyung. The oxidized foils were very fragile, consistent with the embrittlement noted by Daniel et al. The interface was found to have a silicate composition instead of the graphite composition. This is in contrast to the conclusion reached by Daniel et al. that the loss of the graphite layer by oxidation resulted in a clamping stress on the fiber and the resulting brittle "type" fracture. The formation of a silicate that forms a strong bond between the fiber and matrix will accomplish the same result and would also be consistent with the increased interfacial shear stress observed by Daniel et al.

High-temperature mechanical property tests of BMAS–Nicalon composites in air by Sun et al. showed only limited oxidation of near-surface fibers in tests where the stress was below the proportional limit. However, dynamically loaded samples loaded above the proportional limit, or matrix cracking stress, exhibited limited fiber–matrix interface oxidation. Oxygen diffusion along matrix microcracks created by stresses exceeding the proportional limit was thought to be the primary cause for the fiber–matrix interfacial oxidation. This effect was most pronounced under cyclic loading compared to static or quasistatic loading. Embrittlement of a MAS–Nicalon composite during fatigue loading at 500 °C in air was also reported by Heredia et al. A reduction in fatigue life was noted after only 1000 cycles at 500 °C relative to room-temperature tests. Heredia et al. related this loss to a "pest" process where the Nicalon is embrittled and they suggest that the compressive matrix stress in the glass-ceramic matrix– Nicalon composites requires a cyclic stress to reveal this process. Sorensen et al. studied the effect of environment and frequency on the fatigue properties of a CAS–Nicalon composite. They concluded that fiber–interfacial wear processes play a significant role in the loss of fatigue life of these composites and that the environment enhances this wear-induced loss of strength.

Hot Corrosion Environments

High-temperature salt environments will occur in engine components on a Navy gas turbine engine and heat exchangers in coal-fired power plants. There is a strong emphasis on increasing the trust-to-weight ratio of Navy planes and the low-density and high-temperature performance of CMCs are needed to achieve these goals. CAS–Nicalon and LAS–Nicalon composites have been evaluated for this application by Wang et al. They examined the reaction of sodium sulfate with these composites by coating specimens and heating them to 900 °C in either air or argon atmospheres for up to 100 h. The CAS–Nicalon composites exposed in air showed surface cracking and extensive reaction between the salt and the Nicalon fibers. The surface fibers were completely attacked and were totally removed. X-ray diffraction was used to identify the presence of $CaSiO_3$ and $NaAlSiO_4$. The unreinforced CAS glass exposed to the same conditions reacted to form $NaAlSiO_4$ but not $CaSiO_3$. Therefore, the SiC fibers contributed to the reaction products and altered the corrosion reaction. The tensile strength and strain to failure of the CAS–Nicalon composite exposed to sodium sulfate in air was reduced relative to the as-received properties and those for material annealed at 900 °C for 100 h but without the presence of the salt. However, the properties of material exposed to salt in an argon atmosphere showed no degradation in properties. In contrast, the LAS–Nicalon composites did not form additional phases, although there were surface cracks and interdiffusion of Na into the composite and Mg outward diffusion. A 30% strength reduction was noted for the LAS–Nicalon composite exposed to the salt, presumably a result of the surface cracking and Na and Mg interdiffusion.

A thermodynamic evaluation was conducted by Kowalik et al. for the CAS–Nicalon composite exposed to sodium sulfate at 900 °C, as reported by Wang et al. This study suggested the following reaction path: (1) SiC oxidizes to form SiO_2, (2) the silica reacts with the Na_2O in Na_2SO_4 ($Na_2O \cdot SO_3$), the result of reaction 2 may lead to a liquid oxide (or soda slag) phase which may attack the CAS matrix, the SO_3 from reaction 2 may combine with the CaO in the matrix to form $CaSO_4$, and this lost CaO from the matrix is replaced by Na_2O to yield $NaAlSi_3O_8$. These thermodynamic predictions closely match the experimental results reported by Wang et al. Step 1 shows the significance of oxidation in the high-temperature corrosion of these materials.

High-temperature corrosion studies of CAS–Nicalon and BMAS–Tyranno exposed to sodium and magnesium salts have also been conducted by Scott et al. The environments were 3.5% NaCl, 3.5% magnesium salts, and a mix of 3.5% of both sodium and magnesium salts. The samples were coated with these solutions by immersion and then heated to 600 °C, 800 °C, or 1100 °C for up to 60 h. Reaction occurred primarily between the Ca and Mg ions, and the Nicalon in the CAS–Nicalon composite, but the Na ions penetrated the glassy phase and lowered its viscosity in the BMAS–Tyranno composites.

Processing of Ceramic Matrix Composites

Processing Techniques

A number of theories are available for synthesis of various ceramic matrix composites. Structural, thermal and mechanical properties depend on the processing route used for synthesis of ceramic matrix composite material. Therefore, selection of processing route plays an important role for achieving desired properties. A number of conventional and advanced processing routes have been used by various researchers for the fabrication of different kind of CMCs.

Solid State Route

Solid state route is a conventional route to obtain different types of ceramic matrix composites. In this route, different oxide powders are mixed in a stoichiometric proportion to get the required composition. Prepared mixture of powders is kept for wet milling in a ball mill. This is done for few hours to make a homogenous mixture. Grinding media is generally used during mixing to achieve the fine and homogenous powder mixture and the ball to powder ratio and milling duration depends upon the required size of powder. Wet milling is preferred instead of dry milling for better homogeneity which is done in the presence of any liquid medium like propanol. Prepared mixed powder is dried and calcined at a required temperature for moisture and organic removal as well as phase transition. The synthesized powder is mixed with a very small amount of binder and compacted at certain pressure to obtain a desired shape. The prepared pellet is finally dried and then sintered at a high temperature with a soaking period of few hours for phase transformation, microstructure development and densification. Sintering is a process of heating at high temperature to provide bonding and strength to the material which is taken generally 70-90% of the melting temperature.

Sol-gel

In this route, two or more precursor solutions are mixed together in a stoichiometric ratio. This solution is stabilized at room temperature along with continuous stirring. Ammonium hydroxide solution is added drop wise in this stirred solution. As a result of hydro-gel formation, the viscosity of solution increases continuously. pH is finally set to a value of 7-8 at a sufficiently high viscosity. Received gel is kept for ageing at room temperature for few hours. Formed precipitate is washed off by using distilled water. Washing is done for the removal of remaining acidic and basic ions, the filtered precipitate is dried at a temperature below 100°C for few hours. The dried gel is calcined at higher temperature with a hold period of few hours. Achieved powder is then investigated by using different characterization techniques.

Pechini Method

In this route, metal isopropoxide is mixed with nitric acid to form metal nitrate solution. Pechini method involves the formation of polybasic quelates between α-hydroxy carboxylic and metallic ions coming from metal nitrate nona-hydrate. pH of the final solution is maintained to a lower value of approximately 2. Polyesterification of the solution is done on heating below 100 °C and stirring with a poly functional alcohol.

Further heating to a higher temperature at around 175 °C produces a resin having high viscosity. This viscous resin is then transformed into a glassy gel which is rigid and transparent in nature. Attrition milling of this dehydrated gel is done for few hours to convert it in powder form. Sieving of the milled powder is done to achieve the required particle size of powder. The precursor oxide can be finally obtained by heating it at higher temperature.

Laser Synthesis

Laser synthesis includes the non-equilibrium vapor/plasma conditions created at the surface by intense laser pulse. In laser synthesis route, a powder mixture having the combination of different oxides is compacted in pellet form under certain load. The laser treatment of this compact is done by using continuous-action laser radiation (λ = 1064 nm). As a result of irradiation, concave channels (tracks) forms on the surfaces of specimens. Their formation is associated with high temperatures in the irradiation zone, due to which the sintering and melting takes place followed by solidification processes. These tracks were easily removed from the compacts. The lower sides of the tracks were cleaned from the loose slightly sintered powder. Phase formation also occurs during irradiation process. Synthesized products are investigated using different characterization techniques. Figure shows an electron micrograph of cross section of channels formed after laser irradiation.

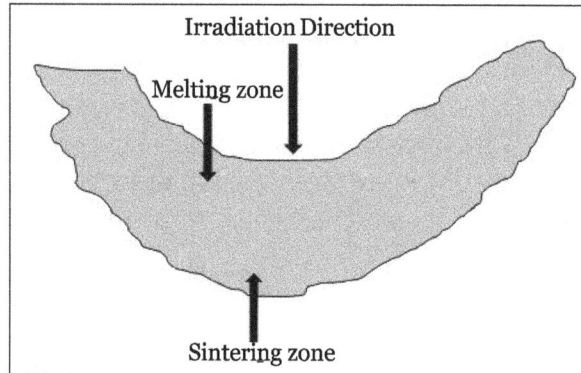

Electron micrographs of cross section of channels formed after laser irradiation.

Melt Synthesis

Melt synthesis route includes the weighing of initial oxide powders and their mixing in ethanol medium. Mixture is dried followed by melting using an arc furnace. Small amount of graphite is added for radiation absorption during melting. A second melting is again done to ensure complete melting which is then cooled to room temperature. This is grounded to convert it in a powder form. Further investigations were done with the help of various characterization techniques.

Co-precipitation Route

In co-precipitation route, different metal salts are mixed homogenously and then precipitated with the help of any basic medium, generally using ammonia solution. Prepared precipitate is then washed to remove extra acid and basic ions. Drying, calcination and then grinding of the washed product is done to achieve powder. Coprecipitation can be done by using two different routes. First is direct precipitation route in which the precipitator is added and mixed homogenously to the salt solution and second is inverse precipitation route in which salt solution is poured and stirred into the precipitator to get precipitate. In second method, the formed precipitate is more finely dispersed than the precipitate by first method. Co-precipitation method offers following advantages. They are:

1. Simple and rapid preparation method.

2. Possibility of controlling particle size and composition easily.

3. Various possibilities to modify the particle surface state and overall homogeneity.

4. Smallest size particles with rounded shape can be obtained by using co-precipitation route.

Hydrothermal Synthesis

In hydrothermal route, fine powders can be synthesized by using initial precursor powder. These precursors are charged into a teflon-lined autoclave. The reactor is heated to a temperature in the range of approximately 100-150 °C under some uniaxial pressure for different reaction times. Obtained powder is washed with distilled water to remove soluble part which is dried in hot air oven for few hours for complete removal of hygroscopic water. Sintering of the dried powder is done at a high temperature for few hours. The control of pH as well as solubility is an important aspect for hydrothermal synthesis. It is reported that the increase in pH value helps in reducing the particle size as well as de-agglomeration of the powder. Small size particles which have higher solubility, tends to form large crystals during sintering. This change in morphology helps in densification and thus the improvement in mechanical behavior occurs. Recent advancement is done by using a polymer as a surfactant during hydrothermal synthesis route.

The combination of hydrothermal route with the use of surfactant helps in achieving following advantages:

- Particle size of the synthesized powder is in nano range (< 100 nm).
- Homogenous and fine grained microstructure of finally sintered product.
- Reduced synthesis temperature and short reaction time.
- Well crystallized solid solution which avoids further thermal treatments.

Fabrication of Ceramic Matrix Composites by Polymer Infiltration and Pyrolysis

Polymer Infiltration and Pyrolysis (PIP) is the method of fabrication of Ceramic Matrix Composites comprising an infiltration of a low viscosity polymer into the reinforcing ceramic structure (e.g. fabric) followed by pyrolysis: heating the polymer precursor in the absence of oxygen when it decomposes and converts into a ceramic. The Ceramics produced from polymers by pyrolysis are called polymer derived ceramics:

- Preceramic polymers.
- Polymer Infiltration and Pyrolysis (PIP) process.
- Advantages and disadvantages of Polymer Infiltration and Pyrolysis (PIP).

Preceramic Polymers

Preceramic polymers (polymer precursors) are the Polymers, which can be converted into Ceramics by pyrolysis. Molecules of preceramic polymers are commonly contain carbon (C) and/ or silicon (Si) but may also contain nitrogen (N), oxygen (O), boron (B), aluminum (Al), titanium (Ti). Polymer Infiltration and Pyrolysis (PIP) technique is used mainly for fabrication Composites with silicon carbide (SiC) matrices from polycarbosilanes (silicon derived polymer precursors): polymethylsilane (PMS) and allhydridopolycarbosilane. The yield of SiC of the precursors is about 65%. Polysilazane may be converted into SiCN or Si_3N_4 with ceramic yield up to 90%. Carbon matrices composites are fabricated by pyrolysis of either carbon thermosetting resins (phenolics, ruran resin, oxidized polystyrene, polyvinyl alcohol) or thermoplastic resins (pitches or coal tar). The carbon yield of these resins is 50-60%.

Polymer Infiltration and Pyrolysis (PIP) Process

- Fabrication of pre-impregnated material (prepreg): The reinforcing fibers are impregnated with a resin and then dried or cured to B-stage (partial curing). In such condition the viscosity of the polymer is increased and the prepreg may be shaped (laid-up).

- Lay-up: The prepreg is shaped by a tooling (mold).

- Molding: The laid-up prepreg is molded. Various molding methods may be used. In the bag molding a rigid lower mold is combined with a flexible upper mold (bag), which is pressed against the prepreg by either atmospheric pressure (vacuum bag mold) or increased air pressure (gas pressure bag mold). The pressurized preform is cured in an autoclave. A combination of a pressure with an increased temperature may also be achieved in compression molding.

- Infiltration of a preceramic polymer: The pores of the reinforcing structure are filled with a low viscosity solution of a preceramic polymer when the preform is immersed into it. The infiltration process is driven by the capillary forces therefore it is commonly conducted at normal pressure, however it may also be vacuum- or pressure-assisted.

- Pyrolysis: Pyrolytic decomposition of the preceramic polymer is performed in the atmosphere of Argon at a temperature in the range 1472-2372 °F (800-1300 °C). Nitride matrices (e.g. silicon nitride) are fabricated in the atmosphere of Nitrogen (N_2) or Ammonia (NH_3). Volatile products such as CO, Hydrogen (H_2), CO_2, CH_2, H_2O are released as a result of pyrolysis forming a porous structure of the resulting ceramic matrix. The value of the ceramic yield is determined by the weight loss (amount of the released volatiles).

- Multiple re-infiltration and pyrolysis: The infiltration-pyrolysis cycle is repeated 4-10 times in order to decrease the residual porosity of the ceramic matrix.

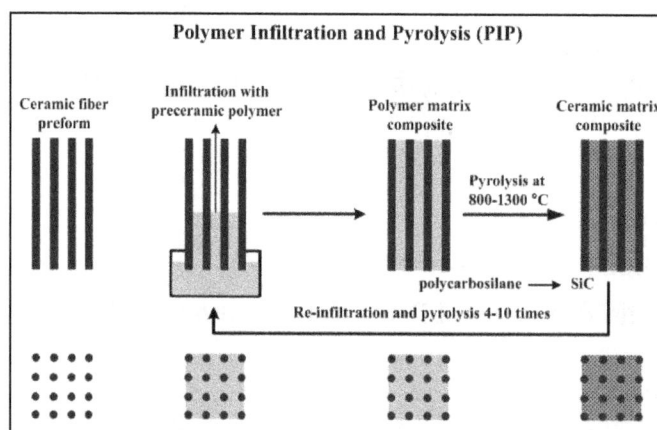

Polymer Infiltration and Pyrolysis (PIP)

Advantages and Disadvantages of Polymer Infiltration and Pyrolysis

Advantages of Polymer Infiltration and Pyrolysis (PIP):

- Fibers damage is prevented due to the processing at a relatively low temperature;
- Good control of the matrix composition and the microstructure;
- Reinforcing phase of different types (particulate, fibrous) may be used;
- Net shape parts may be fabricated;

- Matrices of various compositions (silicon carbide, silicon nitride, silicon carbonitride) may be obtained;
- No residual silicon is present in the matrix.

The disadvantages of the Polymer Infiltration and Pyrolysis (PIP):

- The fabrication time is relatively long due to the multiple infiltration-pyrolysis cycle;
- There is a residual porosity decreasing the mechanical properties of the composite;
- Relatively high production cost (higher than in Liquid Silicon Infiltration method).

Fabrication of Ceramic Matrix Composites by Chemical Vapor Infiltration

- Chemical Vapor Infiltration method of Ceramic Matrix Composites fabrication is a process, in which reactant gases diffuse into an isothermal porous preform made of long continuous fibers and form a deposition. Deposited material is a result of chemical reaction occurring on the fibers surface.

- The infiltration of the gaseous precursor into the reinforcing ceramic continuous fiber structure (preform) is driven by either diffusion process or an imposed external pressure.

- The deposition fills the space between the fibers, forming composite material in which matrix is the deposited material and dispersed phase is the fibers of the preform.

- Chemical Vapor Infiltration (CVI) is similar to Chemical Vapor Deposition (CVD), in which deposition the forms when the reactant gases react on the outer substrate surface.

- Chemical Vapor Infiltration is widely used for fabrication of silicon carbide matrix composites reinforced by silicon carbide long (continuous) fibers.

- Commonly the vapor reagent is supplied to the preform in a stream of a carrier gas (H2, Ar, He). Silicon carbide (SiC) matrix is formed from a mixture of methyltrichlorosilane (MTS) as the precursor and Hydrogen as the carrier gas. Methyltrichlorosilane is decomposed according to the reaction:

$$CH_3Cl_3Si \rightarrow SiC + 3HCl$$

The gaseous hydrogen chloride (HCl) is removed from the preform by the diffusion or forced out by the carrier stream.

- Carbon matrix is formed from a methane precursor (CH_4).

- The ceramic deposition is continuously growing as long as the diffusing vapor is reaching the reaction surface.

- The porosity of the material is decreasing being filled with the formed solid ceramic. However in the course of the CVI process the accessibility of the inner spaces of the preform is getting more difficult due to filling the vapor paths with the forming ceramic matrix. The precursor transportation is slowing down. The growing solid phase separates the spaces in the material from the percolating network of the vapor precursor. Such inaccessible pores do not decrease any longer forming the residual porosity of the composite.

- The matrix densification stops when the preform surface pores are closed. The final residual porosity of the ceramic composites fabricated by CVI method may reach 10-15%.

Types of Chemical Vapor Infiltration Process

Different versions of Chemical Vapor Infiltration process are classified into five types:

- Isothermal/isobaric (I-CVI) is the most commonly used type of CVI process. The fiber preform infiltrated in I-CVI process has no temperature gradient (kept at a uniform temperature). The reactant gas is supplied to the preform at a uniform pressure (no pressure gradient). I-CVI is a very slow process because of the low diffusion rate.

- Temperature gradient (TG-CVI): In this process the preform is kept at a temperature gradient. The vapor precursor diffuses through the preform from the cooler surface to the hotter inside regions. The temperature gradient enhances the gas diffusion. The precursor decomposes mostly in the hot inner regions since the rate of the chemical reaction is greater at higher temperatures. TG-CVI method allows better densification of the ceramic matrix due to prevention of early closing the surface pores.

- Isothermal-forced flow (IF-CVI) utilizes forced flow (pressure gradient) of the gas precursor penetrating into the uniformly heated preform. The rate of the ceramic matrix deposition is increased by the enhanced infiltration of the forced reactant gas.

- Thermal gradient-forced flow (F-CVI) combines the effects of the both temperature gradient and forced flow (pressure gradient) enhancing the infiltration of the vapor precursor. A scheme of Chemical Vapor Infiltration process is shown in the picture below. The presented process combines both temperature gradient and pressure gradient for reduction of densification time. Temperature gradient in preform is achieved by heating the top region of it when the bottom region is cooled. Pressure gradient is determined by the difference in the pressures of the entering and exhausting gases.

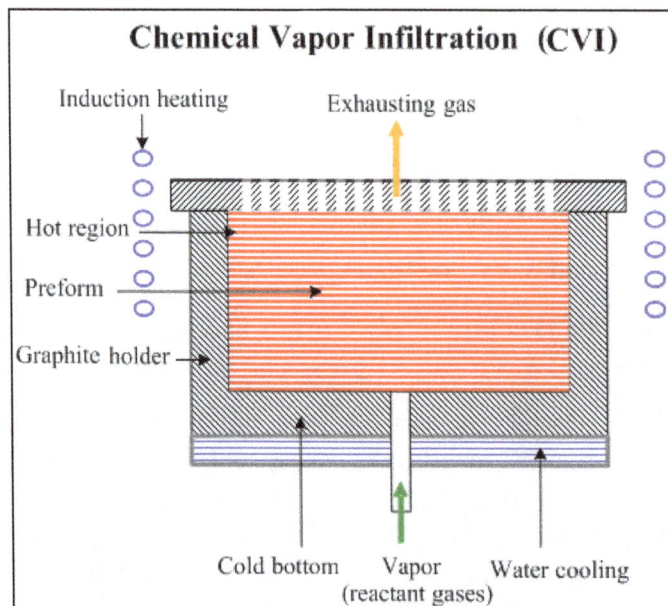

Chemical Vapor Infiltration (CVI)

- Pulsed flow (P-CVI): In P-CVI process the surrounding precursor gas pressure changes rapidly. The pressure changes in each cycle are repeated many times. A cycle of the pressure change consists of the evacuation of the reactor vessel followed by its filling with the reactant gas.

Chemical Vapor Infiltration Process

- Fabrication of the fiber preform.

- Application of a debonding interphase. A thin (commonly 0.1-1 μm) layer of pyrolytic carbon (C) or hexagonal boron nitride (BN) is deposited on the fiber surface by Chemical Vapor Infiltration (CVI) method.

- Infiltration of the preform with a preceramic gaseous precursor. The preform is heated and placed into a reactor with a gaseous precursor. The preform is infiltrated with the gas, which decomposes and forms a ceramic deposit (matrix) on the fiber surface. The process continues until the open porosity on the preform surface is closed.

- Abrading/machining the preform surface in order to open the paths of the percolating network, which allow further densification of the matrix.

- Multiple re-infiltration-abrading cycles until maximum densification is achieved.

- Protection surface coating. The open porosity is sealed in order to prevent a penetration of the environmental gases into the composite during the service. Additional layer protecting the composite surface from the oxidation may be applied over the sealcoat. The coatings are deposited by Chemical Vapor Infiltration (CVI).

Advantages and Disadvantages of Chemical Vapor Infiltration Process

1. Advantages of fabrication of Ceramic Matrix Composites by Chemical Vapor Infiltration (CVI):

   - Low fiber damage due to relatively low infiltration temperatures;

   - Matrices of high purity may be fabricated;

   - Low infiltration temperatures produce low residual mechanical stresses;

   - Enhanced mechanical properties (strength, elongation, toughness);

   - Good thermal shock resistance;

   - Increased Creep and oxidation resistance;

   - Matrices of various compositions may be fabricated (SiC, C, Si_3N_4, BN, B_4C, ZrC, etc.);

   - Interphases may be deposited in-situ.

2. Disadvantages of fabrication of ceramic matrix composites by Chemical Vapor Infiltration (CVI):

   - Slow process rate (may continue up to several weeks);

- High residual porosity (10-15%);

- High capital and production costs.

Fabrication of Ceramic Matrix Composites by Liquid Phase Infiltration

The methods of fabrication of Ceramic Matrix Composites, utilizing infiltration of a liquid into long continuous fibers, are as follows:

- Infiltration of molten ceramic.

- Slurry Infiltration Process (SIP).

- Reactive Melt Infiltration (RMI).

- Polymer Infiltration and Pyrolysis (PIP).

Infiltration of Molten Ceramic

Infiltration of molten ceramic into a fiber preform is limited by low viscosity of molten ceramics and by high temperature causing chemical interaction between the molten matrix and the dispersed phase (fibers). This process (similar to Liquid state fabrication of Metal Matrix Composites) is sometimes used for fabrication glass matrix composites.

Slurry Infiltration Process

Slurry Infiltration Process (SIP) involves the following operations:

- Passing fibers (tow, tape) through a slurry containing particles of the ceramic matrix;

- Winding the fibers infiltrated by the slurry onto a drum and drying;

- Stack of the slurry impregnated fibers in a desired shape;

- Consolidation of the matrix by hot pressing in Graphite die at high temperature.

Reactive Melt Infiltration Process

Reactive Melt Infiltration Process (RMI) is used primarily for fabrication of silicon carbide (SiC) matrix composites (Fabrication of Ceramic Matrix Composites by Liquid Silicon Infiltration (LSI)). The process involves infiltration of carbon (C) containing preform with molten silicon (Si). Infiltration is usually capillary forced. Carbon of the impregnated preform reacts with liquid silicon, forming silicon carbide (SiC). Resulting matrix consists of silicon carbide and some residual silicon.

When liquid aluminum (Al) is used for infiltration of a preform in oxidizing atmosphere, alumina-aluminum (Al_2O_3 – Al) matrix is formed (Fabrication of Ceramic Matrix Composites by Direct Oxidation Process).

Reactive Melt Infiltration method is fast and relatively cost effective. Materials fabricated by RMI method possess low porosity and high thermal conductivity and electrical conductivity.

Polymer Infiltration and Pyrolysis

Polymer Infiltration and Pyrolysis (PIP) involves the following operations:

- Fiber preform (or powder compact) is soaked with a soft (heated) polymer, forming polymeric precursor.

- The polymer is cured (cross-linked) at 480 °F (250 °C).

- The polymer precursor is then pyrolyzed at 1472-2372 °F (800-1300 °C). As a result of pyrolysis the polymer converts to ceramic. Pyrolysis causes shrinkage of the matrix material and formation of pores (up to 40 vol.%).

- The pyrolyzed polymeric cursor may be hot pressed for densification. Hot pressing is limited by possible damage of fibers.

- Infiltration – pyrolysis cycle is repeated several times until the desired density is achieved.

Matrices consisting of carbon, silicon carbide, silicon oxycarbide, silicon nitride and silicon oxynitride may be fabricated by PIP method.

The following materials are used as polymers in Polymer Infiltration and Pyrolysis method:

- Thermosets (thermosetting resins).

- Pitches or other carbon-containing liquids for fabrication of carbon matrix.

- Polycarbosilane, Polysilastyrol, Dodecamethylcyclohexasilan for fabrication of silicon carbide matrix.

Polymer Infiltration and Pyrolysis method are simple low temperature methods, which allow production of intrinsic parts.

Fabrication of Ceramic Matrix Composites by Liquid Silicon Infiltration

Liquid Silicon Infiltration (LSI) process is a type of Reactive Melt Infiltration (RMI) technique, in which the ceramic matrix forms as a result of chemical interaction between the liquid metal infiltrated into a porous reinforcing preform and the substance (either solid or gaseous) surrounding the melt.

Liquid Silicon Infiltration (LSI) is used for fabrication of silicon carbide (SiC) matrix composites. The process involves infiltration of carbon (C) microporous preform with molten silicon (Si) at a temperature exceeding its melting point 2577 °F (1414 °C).

The liquid silicon wets the surface of the carbon preform. The melt soaks into the porous structure driven by the capillary forces. The melt reacts with carbon forming silicon carbide according to the reaction:

$$Si(liquid) + C(solid) \rightarrow SiC(solid)$$

SiC produced in the reaction fills the preform pores and forms the ceramic matrix. Since the molar volume of SiC is less than the sum of the molar volumes of silicon and carbon by 23%, the soaking of liquid silicon continues in course of the formation of silicon carbide. The initial pore volume fraction providing complete conversion of carbon into silicon carbide is 0.562. If the initial pore volume fraction is lower than 0.562 the infiltration results in entrapping residual free silicon. Commonly at least 5% of residual free silicon is left in silicon carbide matrix.

The porous preform may be fabricated by either pyrolysis of a polymerized resin or by Chemical Vapor Infiltration (CVI). The preform microstructure is important for complete infiltration. Large pores helps to obtain a complete infiltration but may result in non-complete chemical interaction and formation of a structure with high residual free silicon and unreacted carbon. Small preform pores results in more complete chemical reaction but in non-complete infiltration due to the blockage (chock-off) of the infiltration channels.

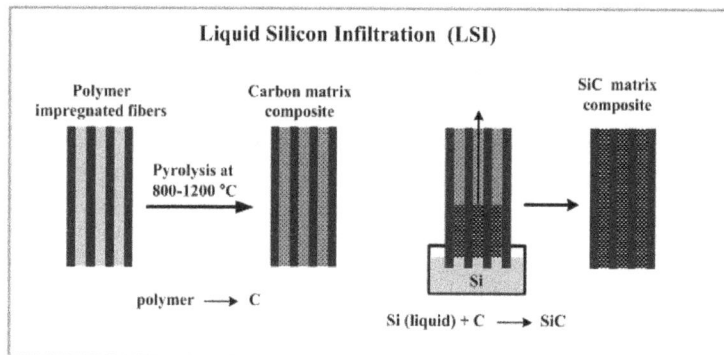

In contrast to the composites fabricated by Polymer Infiltration and Pyrolysis (PIP) and Chemical Vapor Infiltration (CVI) ceramic matrices formed by Liquid Silicon Infiltration are fully dense (have zero or low residual porosity).

The infiltrated at high temperature molten silicon is chemically active and may not only react with the carbon porous preform but also attack the reinforcing phase (SiC or C fibers, whiskers, or particles). A protective barrier coating (interphase) of SiC, C or Si_3N_4 prevents the damage of the fibers by the melt. The barrier coatings are applied over debonding coatings (pyrolytic carbon (C) and hexagonal boron nitride (BN)). The interphases may be deposited by Chemical Vapor Infiltration (CVI). The protective barrier from pyrolytic carbon is formed by Polymer Infiltration and Pyrolysis (PIP).

Liquid Silicon Infiltration Process

- Application of Interphases: A thin (commonly 0.1-1 μm) layer of a debonding phase (pyrolytic carbon (C) or hexagonal boron nitride (BN)) is deposited on the fiber surface by Chemical Vapor Infiltration (CVI) method. In addition to this the fibers are protected from the highly reactive liquid silicon by a barrier coating (commonly SiC). The interphases are deposited by Chemical Vapor Infiltration (CVI).

- Fabrication of the prepreg: The reinforcing fibers (tow, tape, weave) are impregnated with a resin and then dried or cured to B-stage (partial curing). The resin contains carbon, which further will react with molten silicon.

- Lay-up: The prepreg is shaped by a tooling (mold).

- Molding: The laid-up prepreg is molded. Various molding methods may be used. In the bag molding a rigid lower mold is combined with a flexible upper mold (bag), which is pressed against the prepreg by either atmospheric pressure (vacuum bag mold) or increased air pressure (gas pressure bag mold). The pressurized preform is cured in an autoclave. A combination of a pressure with an increased temperature may also be achieved in compression molding.

- Pyrolysis: Pyrolytic decomposition of the preceramic polymer is performed in the atmosphere of Argon at a temperature in the range 1472-2192 °F (800-1200 °C). Volatile products are released as a result of pyrolysis forming a porous carbon structure.

- Primary machining: This operation may be performed after the steps of molding and/or pyrolysis.

- Infiltration of the porous prepreg with Liquid Silicon: The prepreg is immersed into a furnace with molten silicon where its porous carbon structure is infiltrated with the melt. The infiltration process is driven by the capillary forces. Liquid silicon reacts with carbon forming in situ silicon carbide matrix.

- Final machining.

Advantages and Disadvantages of Liquid Silicon Infiltration (LSI) Process

1. Advantages of fabrication of Ceramic Matrix Composites by Liquid Silicon Infiltration (LSI):

   - Low cost.

   - Short production time.

   - Very low residual porosity.

   - High thermal conductivity.

   - High electrical conductivity.

   - Complex and near-net shapes may be fabricated.

2. Disadvantages of fabrication of ceramic matrix composites by Liquid Silicon Infiltration (LSI):

   - High temperature of molten silicon may cause a damage of the fibers.

   - Residual silicon is present in the carbide matrix.

   - Lower mechanical properties of the resulting composite: strength, modulus of elasticity.

Fabrication of Ceramic Matrix Composites by Direct Oxidation Process

- Direct metal oxidation process (Dimox) of Ceramic Matrix Composites fabrication is a type of Reactive Melt Infiltration (RMI) technique, involving a formation of the matrix in the reaction of a molten metal with an oxidizing gas.

- Preform of dispersed phase (fibers, particles) is placed on the surface of parent molten metal in an atmosphere of oxidizing agent (Oxygen).

- Two conditions are necessary for conducting direct oxidation process: dispersed phase is wetted by the melt; dispersed phase does not oxidize in an atmosphere of oxygen. Liquid metal oxidizes when it is in contact with oxygen, forming a thin layer of ceramic with some dispersed phase incorporated in it.

- Capillary effect forces the melt to penetrate through the porous ceramic layer to the reaction front where the metal reacts with the gas resulting in growing the ceramic matrix layer.

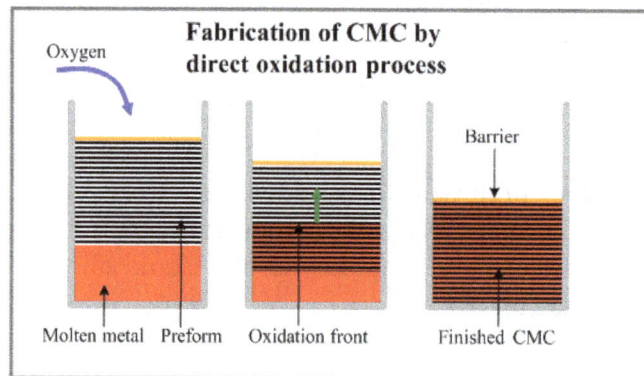

Fabrication of CMC by direct oxidation process

Oxygen

Barrier

Molten metal   Preform   Oxidation front   Finished CMC

- The melt advances to the reaction front continuously at a rate limited by the oxidation reaction rate.

- Some residual metal (about 5-15% of the material volume) remains in the inter-granular spaces of the ceramic matrix.

- The resulting materials have no pores and impurities, which are usually present in ceramics fabricated by sintering (binders, plasticizers, lubricants, deflocculants, water etc.).

- Commonly Direct Melt Oxidation (DIMOX) technique is used for fabrication composites with the matrix from aluminum oxide (Al_2O_3). A reinforcing preform (SiC or Al_2O_3 in either particulate or fibrous form) is infiltrated with a molten aluminum alloy heated in a furnace to a temperature 1652-2102 °F (900-1150 °C).

- The aluminum alloy is doped with additives (e.g. magnesium, silicon) improving the wettability of the reinforcing phase with the melt and enhancing the oxidation process.

- The typical rate of DIMOX process is 0.04-0.06/hr (1-1.5 mm/hr). In principle the direct oxidation process and the oxide growth may continue even after the reaction front has reached the outer surface of the preform. The aluminum oxide will be deposited over the preform changing its dimensions. In order to prevent an advance of the reaction front beyond the preform surface it is coated with a gas permeable barrier. The ceramic matrix growth stops when the reaction front reaches the barrier.

Direct Metal Oxidation Process

- Lay-up: At the lay-up stage the fibrous preform is shaped.

- Application of Interphases: A thin (commonly 0.1-1 μm) layer of a debonding phase (pyrolytic carbon (C) or hexagonal boron nitride (BN)) is deposited on the fiber surface by Chemical Vapor Infiltration (CVI) method.

- Deposition of a gas permeable barrier on the preform surface: The surface through which the melt should wick into the preform is not coated.

- Direct Metal Oxidation: The preform is put in contact with liquid aluminum alloy. The melt wicks into the reinforcing structure through the non-coated surface. The oxidant (air) penetrates into the preform in the opposite direction through the gas permeable barrier. Aluminum and oxygen meet at the reaction front and form the growing layer of the oxide matrix. The process terminates when the reaction front reaches the barrier coating.

- Removal of excessive aluminum: The residual aluminum is removed from the part surface.

Advantages and Disadvantages of Direct Metal Oxidation Process

1. Advantages of DIMOX process:

   - Low shrinkage: Near-net shape parts may be fabricated;

   - Inexpensive and simple equipment;

   - Inexpensive raw materials;

   - Good mechanical properties at high temperatures (e.g. creep strength) due to the absence of impurities or sintering aids;

   - Low residual porosity.

2. The disadvantages of DIMOX process:

   - Low productivity – growth rate is about 0.04/hour (1 mm/hour). The fabrication time is too long: 2-3 days.

   - Residual (non-reacted) aluminum may be present in the oxide matrix.

Fabrication of Ceramic Matrix Composites by Slurry Infiltration

Slurry is a dispersion of ceramic particles in a liquid carrier, which may also contain additives such as binders and wetting agents.

Slurry Infiltration method of fabrication of Ceramic Matrix Composites utilizes a slurry percolating into a porous reinforcing preform. The infiltration process is driven by the capillary forces. After the infiltration process has completed, the preform is dried and hot pressed forming a ceramic matrix composite.

- Ceramic matrices produced by Slurry Infiltration.

- Slurry Infiltration process.

- Advantages and disadvantages of Slurry Infiltration.

Ceramic Matrices Produced by Slurry Infiltration

Slurry Infiltration is used for fabrication of ceramic, ceramic-glass and glass matrices:

- Alumina (AL_2O_3);

- Silica (SiO_2);

- Glass;

- Mullite ($3AL_2O_3\ 2SiO_2$);

- Silicon carbide (SiC);

- Silicon nitride (Si_3N_4).

The method of fabrication of ceramic matrix composites by Slurry Infiltration technique is similar to Sol-Gel Infiltration. However Slurry Infiltration produces denser structure with smaller shrinkage during processing due to higher content of solids. Pressure or vacuum assisted Slurry Infiltration allows further increase of the density of the resulting ceramic composite.

Slurry Infiltration Process

- Slurry infiltration: The reinforcing fibers passe through a slurry which penetrates into the porous structure of the reinforcing phase. The driving force of the infiltration is capillary effect but the process may be enhanced by vacuum or pressure.

- Lay-up: The prepreg (infiltrated fibers) is wound onto a mandrel. Then it is dried, cut and laid-up. After drying they are cut and laid-up on a tooling (mold).

- Hot pressing: Hot pressing (sintering, densification) is performed at high temperature and increased pressure, which enhance the diffusion of the ceramic material between the particles incorporated into the fibers structure. The particles consolidate resulting in a low porosity densified composite.

Advantages and Disadvantages of Slurry Infiltration

1. Advantages of Slurry Infiltration:

   - Low porosity;

   - Good mechanical properties.

2. Disadvantages of Slurry Infiltration:

   - The reinforcing fibers may be damaged by the high pressure applied in the hot pressing stage.

   - Hot pressing operation requires relatively expensive equipment;

   - Relatively small and simple parts may be fabricated.

Fabrication of Ceramic Matrix Composites by Sol-gel Process

- Sol-Gel process of a fabrication of Ceramic Matrix Composites involves preparation of the matrix from a liquid colloidal suspension of fine ceramic particles (sol), which soaks a pre-form and then transforms to solid (gel).

- Colloidal suspension is formed as a result of chemical reaction when very small particles with radii up to 100 nm (nanoparticles) precipitate within a liquid (water or organic solvent).

- Liquid sols have a low viscosity therefore they easily infiltrate into the preform.

- At elevated temperatures sols containing organometallic compounds (e.g. alkoxides) undergo cross-linking (polymerization) by either the polycondensation or hydrolysis mechanism.

- Polymerization converts sol into gel – a polymer structure containing liquid. Gels may be transformed into Ceramics at relatively low temperature, which reduces the probability of the reinforcing fiber damage.

- Alumina matrix ceramic composites may be prepared from alumina gel, which forms in hydrolysis (decomposition as a result of reaction with water) of aluminum alkooxides. Since the amount of ceramic in gels is relatively low they undergo significant shrinkage after drying. The densification of the ceramic matrix is commonly increased by repeating the infiltration-drying cycle several times until the desired density is achieved. Further increase of the volumetric yield of ceramic of a Sol-Gel may be achieved by an addition of ceramic particles. The added ceramic particles also decrease a formation of cracks in the drying stage.

Sol-gel Infiltration Process

- Fabrication of the repreg: The reinforcing fibrous material is immersed into the sol. The sol wicks into the porous structure of the reinforcing phase. Vacuum/pressure may be applied to assist the infiltration process.

- Lay-up: The prepreg is shaped by a tooling (mold).

- Gellation and drying: The sol is heated to 150 °C (302 °F). It is converted into gel, which is then dried at a temperature up to 400 °C (752 °F). Water, alcohol and organic volatile components are removed from the material.

- Repeated re-infiltration and gelation: The sol infiltration-gelation cycle is repeated several times until the desired densification is achieved.

- Firing: The ceramic matrix is consolidated (sintered) at the firing temperature.

Advantages and Disadvantages of Sol-gel Infiltration

1. Advantages of Sol-Gel Infiltration:

   - Less reinforcing fiber damage due to low processing temperature;

- Controllable matrix composition;

- Low equipment cost;

- Low machining cost due to near-net-shape fabrication;

- Large and complex parts may be fabricated.

2. Disadvantages of Sol-Gel Infiltration:

- Possible matrix cracking because of large shrinkage;

- Multiple infiltration-gelatin cycles are required in order to increase the ceramic yield;

- Low mechanical properties;

- High cost of sols.

Properties of the Ceramic Matrix Composites

Mechanical Properties

The high fracture toughness or crack resistance is a result of the following mechanism: under load the ceramic matrix cracks, like any ceramic material, at an elongation of about 0.05%. In CMCs the embedded fibers bridge these crack. This mechanism works only when the matrix can slide along the fibers, which means that there must be a weak bond between the fibers and matrix. A strong bond would require a very high elongation capability of the fiber bridging the crack, and would result in a brittle fracture, as with conventional ceramics. The production of CMC material with high crack resistance requires a step to weaken this bond between the fibers and matrix. This is achieved by depositing a thin layer of pyrolytic carbon or boron nitride on the fibers, which weakens the bond at the fiber/matrix interface leading to the fiber pull-out at crack surfaces. In oxide-CMCs, the high porosity of the matrix is sufficient to establish the weak bond.

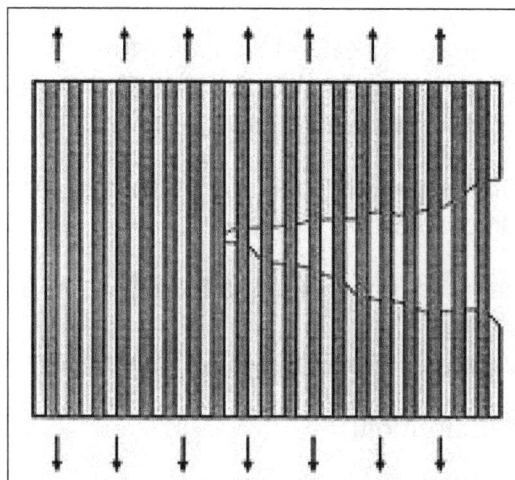

Scheme of crack bridges at the crack tip of ceramic composites.

Thermal and Electrical Properties

The thermal and electrical properties of the composite are a result of its constituents, namely fibers, matrix and pores as well as their composition. The orientation of the fibers yields anisotropic data. Oxide CMCs are very good electrical insulators, and because of their high porosity their thermal insulation is much better than that of conventional oxide ceramics.

The use of carbon fibers increases the electrical conductivity, provided the fibers contact each other and the voltage source. Silicon carbide matrix is a good thermal conductor. Electrically, it is a semiconductor, and its resistance therefore decreases with increasing temperature. Compared to polycrystalline SiC, the amorphous SiC fibers are relatively poor conductors of heat and electricity.

Table: Thermal and Electrical Properties.

Material	CVIC/SiC	LPIC/SiC	LSIC/SiC	CVISiC/SiC	SiSiC
Thermal Conductivity (p)[W/mK	15	11	18	18	>100
Thermal Conductivity (v)[W/mK]	7	5	15	10	>100
Linear Expansion (p)[10-6 .1/K]	13	1.2	0	2.3	4
Linear Expansion (v)[10-6 .1/K]	3	4	3	4	3
Electrical Resistivity (p)[Ω.cm]	-	-	-	-	50

Comments for the table: (p) and (v) refer to data parallel and vertical to fiber orientation of the 2D-fiber structure, respectively. LSI material has the highest thermal conductivity because of its low porosity – an advantage when using it for brake discs. These data are subject to scatter depending on details of the manufacturing processes. Conventional ceramics are very sensitive to thermal stress because of their high Young's modulus and low elongation capability. Temperature differences and low thermal conductivity create locally different elongations, which together with the high Young's modulus generate high stress. This results in cracks, rupture and brittle failure. In CMCs, the fibers bridge the cracks, and the components show no macroscopic damage, even if the matrix has cracked locally. The application of CMCs in brake disks demonstrates the effectiveness of ceramic composite materials under extreme thermal shock conditions.

Applications of Ceramic Matrix Composites

Developments for Applications in Space

During the re-entry phase of space vehicles, the heat shield system is exposed to temperatures above 1500 °C for a few minutes. Only ceramic materials are able to survive such conditions without significant damage and among ceramics only CMCs can adequately handle thermal shocks. The development of CMC-based heat shield systems promises the following advantages:

- Reduced weight.

- Higher load carrying capacity of the system.

- Reusability for several re-entries.

- Better steering during the re-entry phase with CMC flap systems.

NASA-space vehicle X-38 during a test flight.

In these applications the high temperatures preclude the use of oxide fiber CMCs, because under the expected loads the creep would be too high. Amorphous silicon carbide fibers lose their strength due to re-crystallization at temperatures above 1250 °C. Therefore carbon fibers in a silicon carbide matrix (C/SiC) are used in development programs for these applications. The European program HERMES of ESA, started in the 1980s and for financial reasons abandoned in 1992, has produced first results. Several follow-up programs focused on the development, manufacture, and qualification of nose cap, leading edges and steering flaps for the NASA space vehicle X-38.

Application in Brake Disk

Carbon/carbon (C/C) materials have found their way into the disk brakes of racing cars and airplanes, and C/SiC brake disks manufactured by the LSI process were qualified and are commercially available for luxury vehicles. The advantages of these C/SiC disks are:

- Very little wear, resulting in lifetime use for a car with a normal driving load of 300,000 km, is forecast by manufacturers.

- No fading is experienced, even under high load.

- No surface humidity effect on the friction coefficient shows up, as in C/C brake disks.

- The corrosion resistance, for example to the road salt, is much better than for metal disks.

- The disk mass is only 40% of a metal disk. This translates into less unsprung and rotating mass.

The weight reduction improves shock absorber response, road-holding comfort, agility, fuel economy, and thus driving comfort.

The SiC-matrix of LSI has a very low porosity, which protects the carbon fibers quite well. Brake disks do not experience temperatures above 500 °C for more than a few hours in their lifetime. Oxidation is therefore not a problem in this application. The reduction of manufacturing costs will decide the success of this application for middle-class cars.

LeMans prototype DMC disc Φ380x18 mm.

Ferrari F430 front brake disc.

Cement Based Composites

Cement composites are commonly used in construction due to their numerous advantages, including high compressive strength, low manufacturing cost, simple production process, and ease of use. However, they have some basic weaknesses: low tensile strength, low resistance to deformation, and susceptibility to cracking. Ensuring good properties, not only mechanical but also other physico-chemical properties, is a real challenge for present-day material engineers and concrete technologists, particularly in the case of expanded structures, which are often multifactorially prestressed, with complicated shapes and exposed to extreme environmental actions. High-performance and multifunctional cement composites with excellent mechanical properties, durability, and ease of production suitable for the structural material are promising approaches to implementation of cement composite structures in terms of sustainable development.

The greatest and most spectacular achievements in the field of modification of structural materials have been related to the use of nanomaterials. In the construction and building materials industry, nanomaterials have already found commercial application. The investigation by Colston is considered to be one of the first studies of cement composites modified with nanoparticles (NPs). These investigations showed that incorporation of the nanoparticles into cement leads to significant improvements in cement composites' microstructures. Later, certain types of the nanomaterials were discovered to not only improve the resistance to brittle cracking and strength of cement composites, but also give them other properties, thus producing multifunctional composites. The nanomaterials used in the building materials industry include silicon dioxide, titanium dioxide, and zinc oxide. These nanomaterials, together with silver nanoparticles, carbon nanotubes, and nanofibers, are most often used for the production of the commercially available building products containing nano-objects. Only a few nanomaterials have been investigated as additions or admixtures to cement concretes and mortars; these include titanium dioxide, aluminum nano oxide, zinc oxide, nano-$CaCO_3$, and silicon dioxide. Most of the research dealing with the modification of cement-based composites concerns titanium dioxide and silicon dioxide, commonly called nanosilica.

Despite the many available studies on the evaluation of the influence of nanomaterials on the mechanical properties of cement composites, the effects of some nanoparticles have not yet been fully recognized. Among the unrecognized nanomaterials are iron oxides (nano-Fe_2O_3 and nano-Fe_3O_4).

Iron oxide (gamma-Fe_2O_3), in the form of a nanopowder, has been applied in special anticorrosion coatings, silicones, plastics, rubbers, alloys, lithium batteries, lithium iron phosphate batteries, magnetic seals, wear-resistant materials, and targeted drug delivery. Fe_2O_3 is heat-resistant up to a temperature of 450 °C. Above this temperature, the particles lose their magnetic properties. Because the nanoparticle surface area is very large, without surface treatment, the nanoform cannot exist. This means that the nanoparticles cannot produce the nanoproperties. One particle is composed of thousands of single nanoparticles by adsorption of nanoparticles with soft aggregation.

Fe_3O_4 demonstrates high magnetic performance, high saturation magnetization, and low cost. Thus, as a microwave absorption absorber. The main disadvantage of Fe_3O_4 is its poor thermal stability, which may lead to the loss of a single domain pole or the special nature of magnetic materials. Therefore, this property restricts their wide application. The protective coatings of the non-metallic materials, e.g., SiO_2, TiO_2, and Al_2O_3, have been used for improvement of these properties (to increase the thermal stability and diminish oxidation).

Methods of Synthesizing Magnetic Nanostructures

Many studies concerning the synthesis of magnetic nanoparticles were carried out during the recent decades. Significant progress has been achieved in the range of the synthesis, protection, and functionalization of magnetic nanoparticles. The development of such methods as co-precipitation, thermal decomposition and reduction, synthesis of micelles, and hydrothermal synthesis has enabled us to control the size and shape of magnetic nanoparticles. Efficient methods of magnetic protection against corrosion were developed by surface-active (polymer) coating, silica coating, and carbon coating, or by embedding the magnetic nanoparticles inside the matrix or carrier. The unique shapes of magnetic nanoparticles can be obtained by direct synthesis, in which the anisotropic growth is directed by tuning the reaction conditions or by the use of templates, or by the assembling method, in which a high aspect ratio is achieved by assembly from individual building blocks. Methods including co-precipitation, thermal decomposition and/or reduction, micelle synthesis, hydrothermal synthesis, and laser pyrolysis can all be applied for the synthesis of high-quality magnetic nanoparticles. The advantages and disadvantages of the four main metallic nanostructures' synthesis methods are summarized in table. Due to the size and control of the morphology, thermal decomposition seems to be the best synthesis method. Alternatively, microemulsions can be used to synthesize the monodispersion of nanoparticles with various morphologies. The hydrothermal synthesis methods enable the production of high-quality nanoparticles. The co-precipitation and thermal decomposition methods are best known, and enable the production of large amounts of nanomaterial.

Table: A summary comparison of the synthesis methods.

Synthesis method	Degree of complication, conditions	Reaction temperature (°c)	Reaction period	Solvent	Surface-capping agents	Size distribution	Shape control	Yield
Co-precipitation	Very simple, ambient conditions	20–90	Minutes	Water	Needed, added during or after reaction	Relatively narrow	Not good	High/ stable
Thermal decomposition	Complicated, inert atmosphere	100–320	Hours-days	Organic compound	Needed, added during reaction	Relatively narrow	Very good	High/ stable

Microemulsion	Complicated, ambient conditions	20–50	Hours	Organic compound	Needed, added during reaction	Relatively narrow	Good	Low
Hydrothermal synthesis	Simple, high pressure	220	Hours ca. Days	Water-ethanol	Needed, added during reaction	Relatively narrow	Very good	Medium

Concrete, as a basic cement-based construction material, should demonstrate—in addition to the physical and mechanical property requirements—resistance to environmental action. Concrete engineering structures should be durable when exposed to high temperature (fire hazard), low temperature (exploitation in winter), and chemical aggression (corrosion of concrete and steel reinforcement). Due to the poor thermal stability of nano-Fe_3O_4, when applied to cement composites, the nano-SiO_2 shell is used for improving the thermal resistance. In Cendrowski et al, it attempted to synthesize the magnetite-silica structure of the core-shell type for potential use as a modern admixture designed for modification of building cement composites to improve the properties of the cement composites under high temperature and corrosive environments. The nanostructures with a magnetite (Fe_3O_4) core and a solid nano-SiO_2 shell were used. The commercially available nanomagnetite (Sigma-Aldrich, Darmstadt, Germany) with a diameter of 50–100 nm was used as the core in the investigation. The X-ray diffraction (XRD) analysis confirmed the occurrence of the characteristic peaks related to the magnetite's phases.

(a,b) A transmission electron microscopy (TEM) micrograph, (c) A scanning electron microscopy (SEM) micrograph, and (d) The X-ray diffraction (XRD) pattern of nano-Fe_3O_4.

Stöber's method was employed for the synthesis of the silica shell. The thickness of the shell was about 20 nm. After obtaining the nanostructures, the nanomaterial properties were analysed, verifying the efficacy of the synthesis. Then, the chemical stability (in hydrochloric acid) and the thermal stability of the nanocomposites were tested. The transmission electron microscopy (TEM) micrograms of the produced nanostructures are presented in figure.

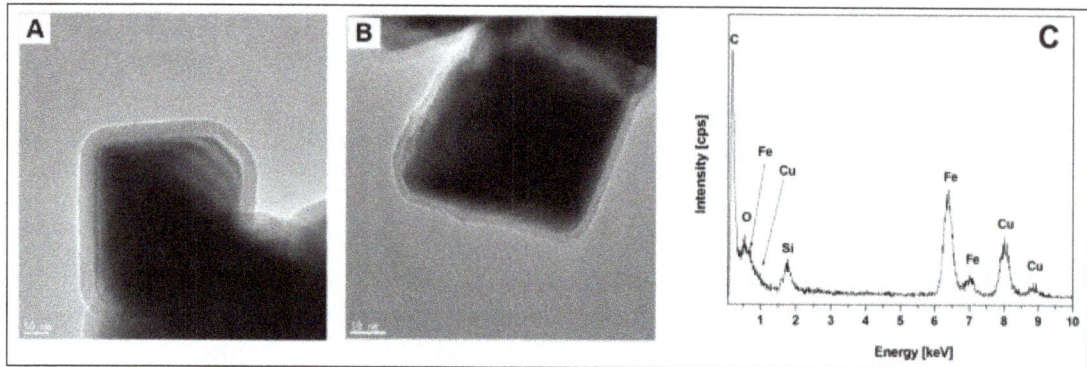

TEM micrographs of (A,B) iron oxide/silica (Fe$_3$O$_4$/SiO$_2$) core-shell structures (Reprinted with permission from. and (C) the energy dispersive X-ray spectrometry (EDS) graph.

The use of a solid shell around the nanomagnetite core prevented the reaction of the acid with the core of the structure, which showed potentially high resistance to aggressive environments.

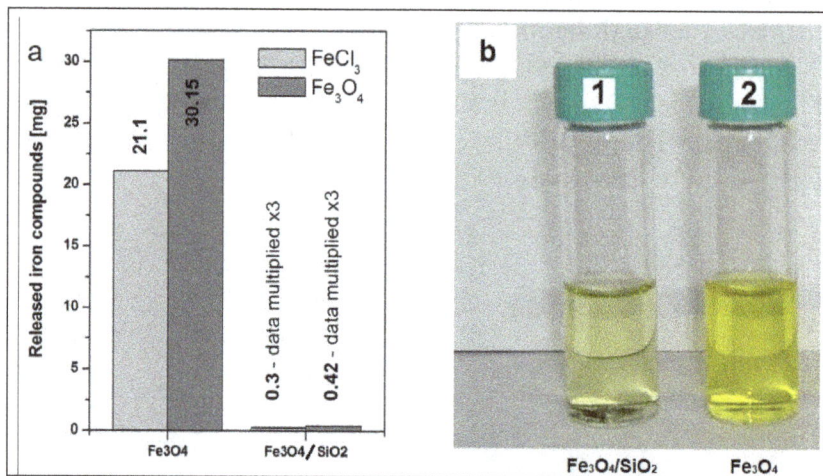

(a) Optical images of the concentrated hydrochloric acid after 24 h dissolution of the Fe$_3$O$_4$/SiO$_2$ and (b) after extraction of undissolved nanomaterials. The samples marked 1 and 2 correspond to the Fe$_3$O$_4$/SiO$_2$ and Fe$_3$O$_4$, respectively. Ultraviolet (UV)-spectroscopy-calculated amounts of the dissolved magnetite and formed iron chloride based on Cendrowski et al.

Thermogravimetric analysis (TGA) showed that, in the case of nanostructures covered with a solid shell, oxidation increased gradually with increasing temperature, which proves the improved resistance of the nanostructure to high temperatures.

The thermal stability of nano-Fe$_3$O$_4$ samples is dependent on many factors, including the nature of the atmosphere medium, the amount of chemical impurities, the size of the crystals, and the temperature to which the material is subjected de Mendonça studied the influence of the SiO$_2$ shell on the thermal stability of nano-Fe$_3$O$_4$; an increase in stability was observed for the shell with an average diameter below 1 nm. Moreover, the MN coating material had a less pronounced change in color (black to gray-reddish-brown), and presented an attraction to a magnet, whereas pure nano-Fe$_3$O$_4$ particles were fully reddish-brown with no magnetic attraction. Other investigations confirmed that the silica layer appears to be effective in protecting magnetite from being converted to other oxide species.

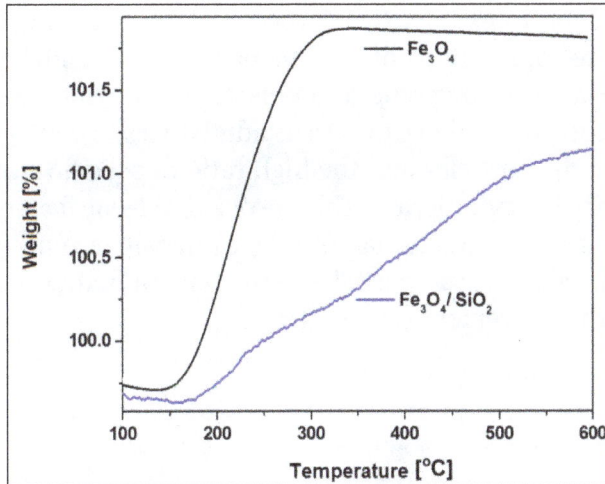

Thermogravimetric analysis of the magnetite nanoparticles (Fe_3O_4) and magnetite nanoparticles with a solid silica shell (Fe_3O_4/SiO_2).

Fe_3O_4/SiO_2 particles have shown resistance to chemically aggressive environments and when exposed to high temperatures. High-resolution transmission electron microscopy (HRTEM) analysis did not show any damage or cracks in the shell of nanostructures heated to 550 °C. The chemical stability of Fe_3O_4/SiO_2 heated to 550 °C showed that heating the solid-silica-coated magnetite nanoparticles had no significant impact on their acid resistance. HRTEM analysis did not reveal any cracks in the solid silica coatings. Therefore, it can be concluded that the thermal stability of the magnetite nanoparticles coated with solid silica is determined by the oxygen diffusion mechanism rather than iron oxide thermal expansion, which could cause shell cracking. The use of the solid shell from the nanosilica enabled the better dispersion of the particles (Fe_3O_4/SiO_2) in the cement matrix.

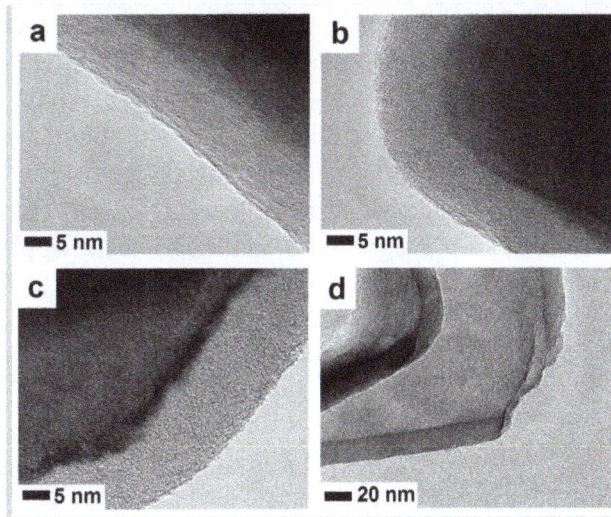

High-resolution transmission electron microscopy (HRTEM) images of the Fe_3O_4/SiO_2 particle (a,b) after exposure to 550 °C and (c,d) after exposure to 550 °C and acid aggression.

Processing of MN-engineered Cementitious Composites

Manufacturing of cement composites with magnetite nanoparticles (MN) covers preparation of the nanomodifier as well as the remaining components of the composite (cement, water, possible chemical admixtures, and additions) and their processing (e.g., mixing/dispersing, molding, and curing), as shown in figure. Application of various nanoparticles, including MN, to the cement composite can cause nanomaterial agglomeration inside the cement matrix. This phenomenon is a key problem related to the practical use of NPs as admixtures in cement composites. Regarding the relatively small size of the particles and the high ratio of specific surface area to volume, the MN show a strong tendency to agglomerate. The use of suitable surfactants or more sophisticated methods of introduction of the nanomaterial into the composite are necessary in such cases. The occurrence of unevenly distributed nanoparticles in the cement matrix negatively affects the properties of the fresh and hardened cement composites.

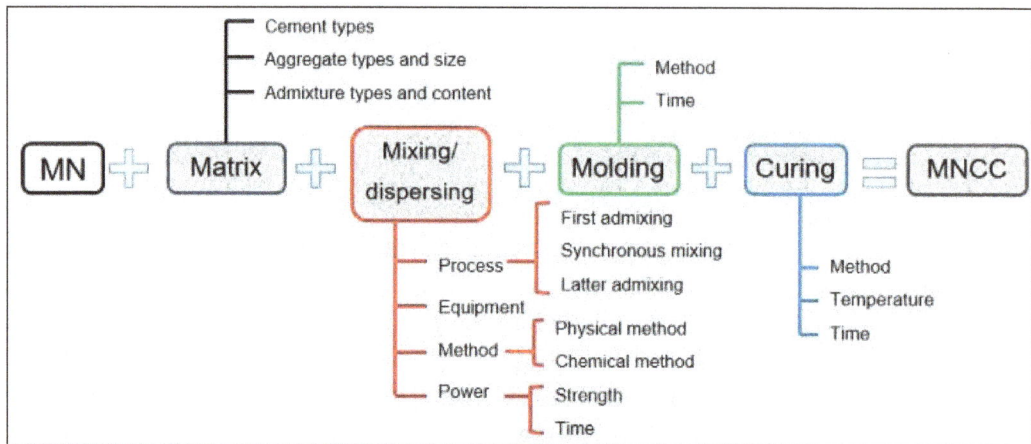

Scheme of the preparation process of magnetite nanoparticles cement composites (MNCC).

Two main forms are employed for the introduction of nanosilica into a cement composite: powder or water suspension. The first method is similar to the method of introducing silica fume into concrete and mainly consists of the mechanical mixing of the cement with the NPs. Even a long mechanical mixing duration cannot ensure the proper dispersion of the nanomaterial. The results described in Kong. show that, during mixing, the agglomerates are partially disintegrated under the action of the shear forces. The process of mixing causes a slight decrease in the number of agglomerates; however, this effect is not full and the particles are not totally broken. Thus, the development of an efficient method of dispersing the nanoparticles in a cement composite is a basic problem for many researchers, posing a significant challenge in the application of nanomaterials in the building materials industry.

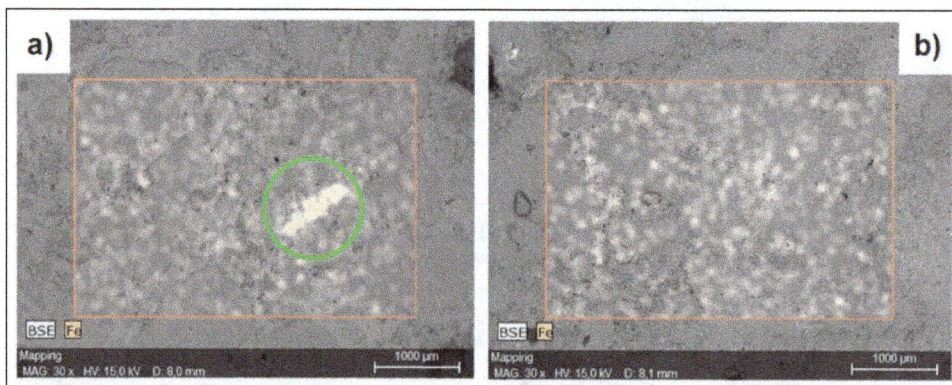

A map of Fe distribution in the cement mortar containing nano-Fe_3O_4: (a) a visible agglomerate of nano-Fe_3O_4 in the cement matrix; (b) a uniform distribution of the nanomaterial in the matrix.

For this reason, researchers more often use the second method to introduce the nanomaterial into the cement composite, involving the dispersion of the nanometric powder in the water before adding it to the dry components of the composite. In this method, ultrasonic dispersion is used in addition to mechanical mixing. The nanoparticles of the magnetic compounds, besides the large specific surface area, are also magnetic, which affects the formation of agglomerates in the cement composite. Sonication appears to be an effective method for dispersing the nanoparticles, including the magnetic NPs. This method, however, is energy-consuming and significantly increases the cost of application.

Figure presents a map of Fe distribution in cement mortars modified with an admixture of nano-Fe_3O_4. Before the introduction of the nanomaterials into the cement mortars, the nano-Fe_3O_4 particles were sonicated in water for 1 min to obtain a uniform dispersion.

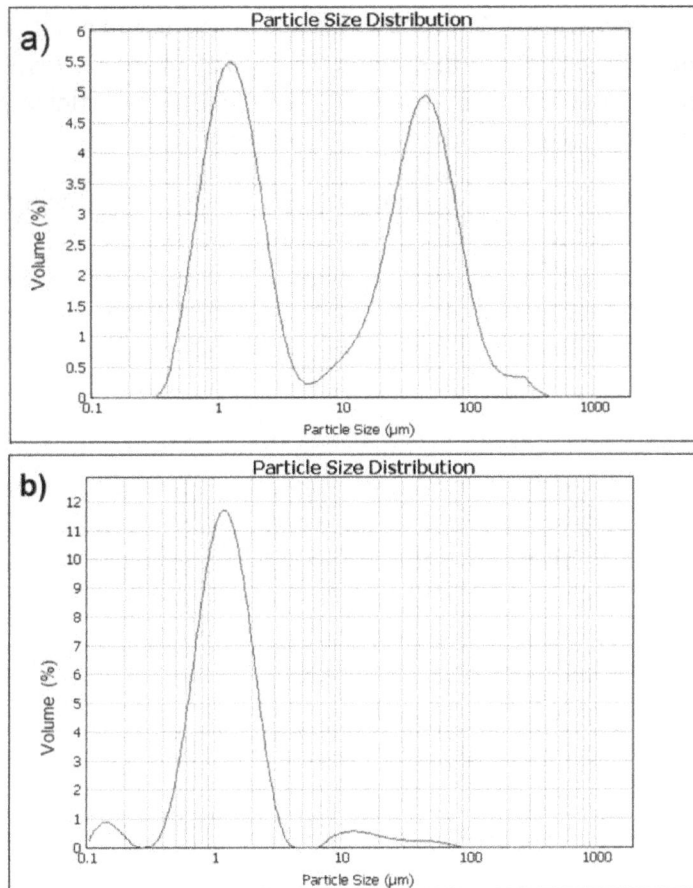

Influence of the ultrasonic dispersion (sonication) on the granulometry of Fe_3O_4 nanoparticles in the solution of water and superplasticizer based on polycarboxylic ether PCE: (a) without sonication and (b) after sonication for 1 min.

The impact of the ultrasonic dispersion (sonication) on the granulometry of the MN in the preparation process of the cement composite is presented in figure. In the first case, the Fe_3O_4 NPs were mixed mechanically with a solution of water and 2% superplasticizer based on polycarboxylic ether (PCE). The agglomerates of Fe_3O_4 particles, which were not broken, are clearly visible in figure. Figure presents the granulometric curve of the same mix after the additional process of sonication for 1 min. Incorrectly prepared dispersion of the nanomaterial and a large number of agglomerates can significantly affect the kinetics of the hydration process and, in consequence, also worsen the structure and final properties of the cement composites. The process described in Singh et al. performed on the nanosilica, showed that the method of introducing the NP into the cement composite significantly influences its rheology and porosity, and thus the mechanical properties of the hardened composite. Incorrect dispersion of the NP in the cement composite can contribute to the creation of local "weak areas" with worsened mechanical features. As was demonstrated by Yang et al, ultrasonic dispersing can cause intensity-dependent (ultrasound frequency) breaking of the nanostructures into particles with smaller diameters.

The most often used admixtures for modern cement composites are water-reducing admixtures (plasticizers) and high-range water-reducing admixtures (superplasticizers). Superplasticizers, particularly often used for high-performance cement composites, are based on the polycarboxylic ethers. The results described in Li et al. confirmed that, in the case of small amounts of nanoparticles in the cement composite (below 1% of the binder mass), the use of a water-reducing admixture mechanically breaks the nanoparticles added to the previously prepared solution of water and water-reducing admixture. For nanomaterials, plasticizers and superplasticizers are surfactants that facilitate the dispersion of the nanomaterial in the composite.

A technique that enables the dispersion of nanoparticles in a water-reducing admixture is not yet available. Many surfactants that are successfully used to disperse nanomaterials, e.g., in polymeric matrices, have been reported to affect the cement hydration kinetics and negatively react with other admixtures. Therefore, plasticizers and superplasticizers (especially polycarboxylic ether (PCE)-based superplasticizers) compatible with cement have been widely tested and evaluated as dispersants.

Another possible solution, particularly in the case of magnetic nanoparticles, seems to be the use of an additional shell of nano-SiO_2. Han demonstrated that TiO_2 covered with a SiO_2 nanoshell can be well-dispersed in a solution of water and water-reducing admixture using mechanical mixing. In the case of the magnetic nanostructure, the use of a nano-SiO_2 shell also weakens the attraction forces between the particles. When large amounts of nanomaterials are used, however, ultrasonic dispersion together with mechanical stirring are necessary for ensuring the uniform distribution of the nanomaterial in the composite. Preparation of such a composite with MNPs was described in Sikora et al. To ensure the better dispersion of MN in the matrix, mechanical mixing and sonication were simultaneously used. The preparation scheme of the composite with MN.

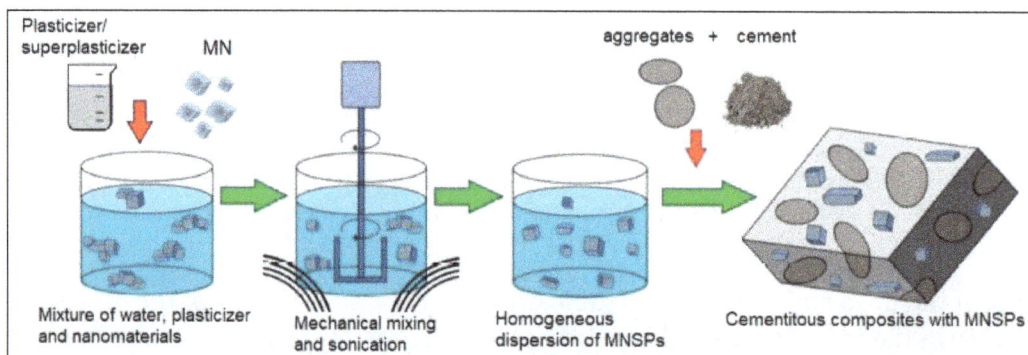

Scheme of the preparation of the cement composite containing magnetite nanoparticles (MN).

The method of composite preparation mainly involves the selection of the method and time of compaction. An incorrect choice of the compaction method can cause additional aeration. Curing of the cement composite specimens containing MN is usually carried out in water or in a chamber with relative humidity above 95%. Additional treatments accelerating cement hydration can be applied, such as the use of hot water. The main methods for manufacturing the cement composites containing MN nanostructures are outlined in.

Table: Processes available for the production of cement composites with MN particles.

Nano-Particles	MN Dispersion Method	Feeding Order	Mixing Method/Time	Molding		Curing	
				Size (mm)		Condition/Temperature	
Fe_3O_4/SiO_2	Shear mixing	W + Sp + NP C + SWN	Stir/3 min Stir/2 min	Vibration/ 50 × 50 × 50 (compressive test)		Lime-saturated water 20 °C	
Fe_3O_4	Shear mixing	C + NP W	Stir/30 min Stir/15 min	—/cylindrical mold: diameter 200, height 300 (compressive test)		———	
Fe_3O_4	Shear mixing	C + NP + M + S + G + W + Sp	———	—/150 × 150 × 150 (compressive test) cylindrical mold: diameter 150, height 300 (indirect tensile test)		Water 20 ± 1 °C.	
Fe_3O_4	Shear mixing	C + NP S + G W + Sp	———	—/150 × 150 × 150 (compressive test) —/cylindrical mold: diameter 150, height 300 (indirect tensile test)		Water 20 ± 1 °C.	
Fe_3O_4	Ultrasonic method + Shear mixing	W + NP + Sp C + S SWN	Stir + ultrasonic/1 min Stir/1 min Stir/2 min	Vibration/ 40 × 40 × 160 (flexural and compressive test)		Water 20 ± 2 °C.	
$Fe_3O_4Fe_3O_4/SiO_2$	Ultrasonic method + Shear mixing	W + NP + Sp C + S SWN	Stir + ultrasonic/30 min Stir/1 min Stir/2 min	Vibration/ 40 × 40 × 160 (flexural and compressive test)		Water 20 ± 2 °C	

Properties of Cement Composites Containing Nano-Fe_3O_4

Hydration Process

The process of cement hydration is controlled not only by the mineral components of the cement particles, the size of the particles (specific surface area), the amount of the added water, and the temperature, but also by the type and content of the used nanoparticles. Studies concerning the influence of the nano-Fe_3O_4 admixture on the process of cement hydration are scarce. Sikora et al. investigated cement pastes (Portland cement 42.5 R) modified with nano-Fe_3O_4 at 5% of the cement mass. The testing of the pastes during the first three days of curing did not show any changes in the hydration process. XRD analysis of the cement pastes containing nano-Fe_3O_4 in amounts from 1% to 5% of the cement mass did not reveal new hydration products in the paste after 7 days of curing The main phases are calcium silicate hydrates (C–S–H), portlandite (CH), and $CaCO_3$. With increased content of nano-Fe_3O_4 particles, the peaks related to the presence of magnetite increase. The XRD obtained after 7 days of curing displayed the same hydrate phases as the reference sample. Likewise, after 28 days of curing, no phase changes occurred. Amin et al.

investigated pastes containing Portland cement and high-slag cement, containing 0.1, 0.3, and 0.5% nano-Fe_3O_4 in relation to the cement mass. The XRD results obtained after 1, 7, and 28 days confirmed that Fe_3O_4 nanoparticles do not affect the rate of the cement hydration nor the characteristics of the hydration products.

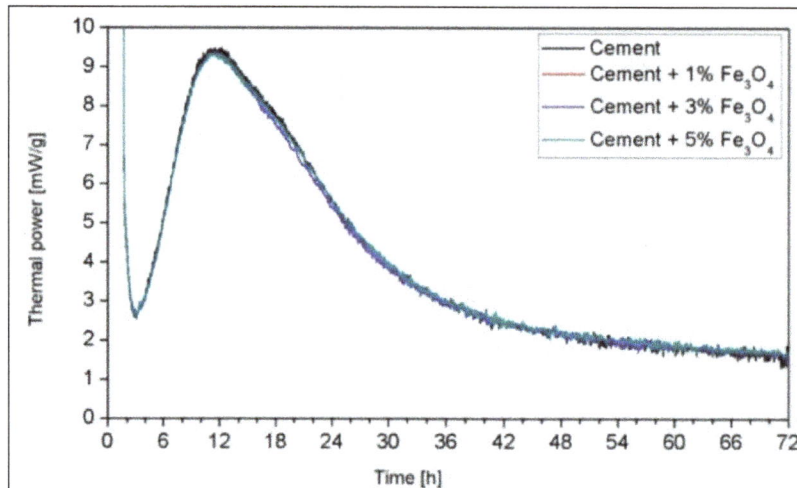

Heat flow calorimetry of the cement paste (0.5 wt %) with different dosages of nano-Fe_3O_4.

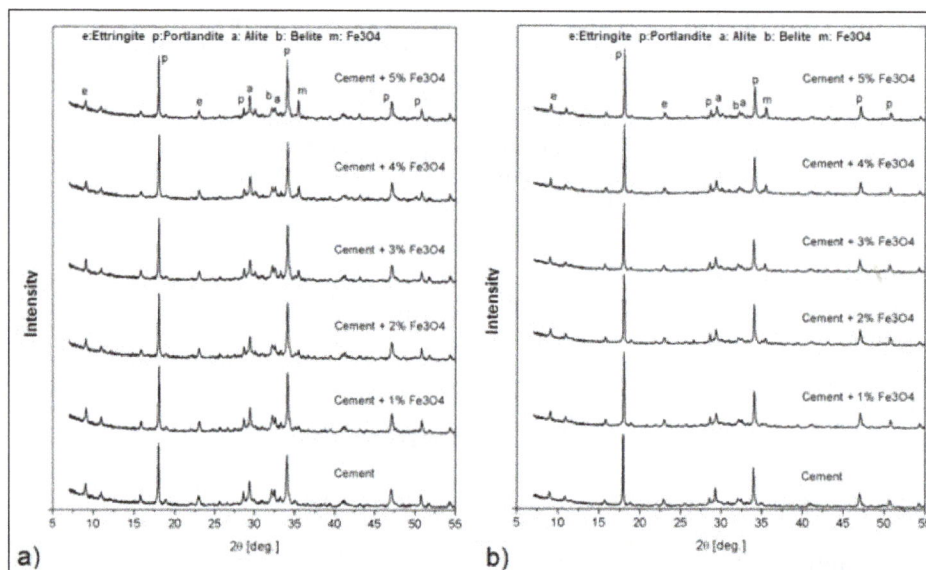

XRD spectra of the cement pastes containing nanomagnetite after (a) 7 and (b) 28 days of curing.

Workability of Composites

Many investigations into cement composites modified with various nanomaterials have verified that the admixture of nanoparticles mostly negatively affects the workability of the fresh composites. Slump and slump flow are often used for the assessment of the workability of cementitious composites. A decrease in the flow with increasing nanoparticles content in the composite has been observed in many studies. This is connected to the small size effect and large specific surface area of the nanostructures. The large specific surface area and the resulting high water demand cause a reduction in the amount of free water available for the hydration process, and consequently lead to the limited workability of the cement composites.

Most investigations into the consistency have been carried out for cement mortars modified with various amounts of MN. These tests are performed by the method according to EN 1015-3. Sikora tested the consistency of cement mortars containing 1–5% nano-Fe_3O_4 (in relation to the cement mass), without superplasticizer (water–cement ratio, $w/c = 0.5$ for all mortars). The test results are presented in figure. The presence of nano-Fe_3O_4 did not significantly influence the consistency of the mortars, despite the nanometric characteristic of the modifier. This is strongly connected to the nonporous morphology of nano-Fe_3O_4 and the more hydrophobic nature of nanomagnetite compared with other nanomaterials, such as nano-SiO_2 or nano-TiO_2, affecting the consistency and worsening the workability of the cement composites.

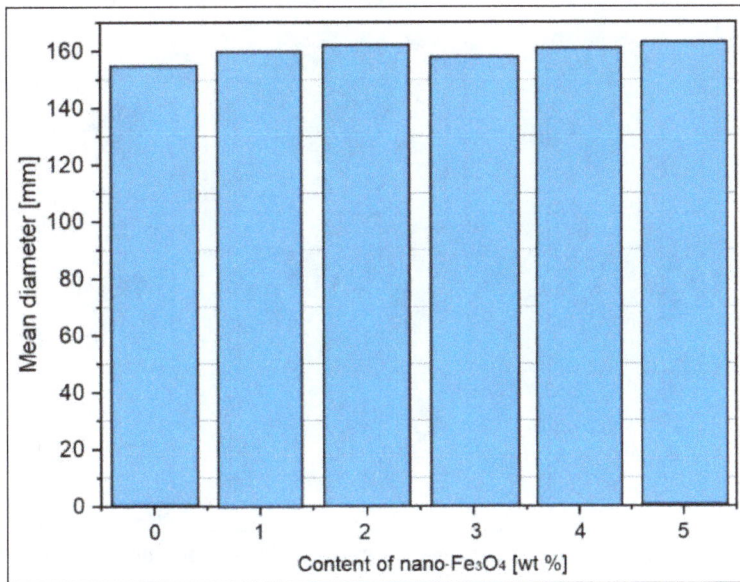

The consistency of the fresh mortars with nano-Fe_3O_4.

Bolhassani and Sayyahmanesh studied the workability of mortars containing 0.05, 0.1, and 0.2% nano-Fe_3O_4. All mortars were designed with a low coefficient $w/c = 0.28$. The superplasticizer was used in amounts proportional to the content of nanomaterial (0.05%, 0.18%, and 0.55% of the binder mass, respectively). The addition of superplasticizer for obtaining a workability of mortar similar to that of the reference (without nanomodifier) appeared necessary only for mortar containing 0.2% nanomaterial.

Structure of Cement-based Composites Modified with MN

The mechanical properties of the cement composites modified with MN depend on the type, content, and form of the hydration products. Nano-Fe_3O_4 particles can accelerate the rate of cement hydration due to their high activity. This phenomenon has been described in Amin et al, where small amounts of Fe_3O_4 nanometric particles were introduced into the cement paste (up to 0.3% of the cement mass). The diffusion of the hydration products starts during the hydration process and the nanometric particles are surrounded and become the crystallization nuclei around which the hydrates focus. If the amount of nanoparticles is optimum, the crystallization process will be controlled and the growth of the $Ca(OH)_2$ crystals will be stopped by the nanoparticles, which in turn will improve the cement paste microstructure. However, when the amount of nanoparticles is too large, the $Ca(OH)_2$ crystals cannot grow sufficiently due to the limited space in the matrix.

This leads to the increase in shrinkage and creep in the cement matrix, resulting in the increased porosity of the matrix. Moreover, the nanoparticles fill the pores due to their nanometric size and the so-called "nanofiller effect" is observed, causing further compaction of the microstructure.

These two phenomena lead to an improvement in the composite's microstructure by decreasing the number of pores, strengthening the bond between the aggregate and cement paste, and increasing the density of the cement composite. The scheme of the compaction process of the composite's structure under the influence of nano-Fe_3O_4 is shown in figure.

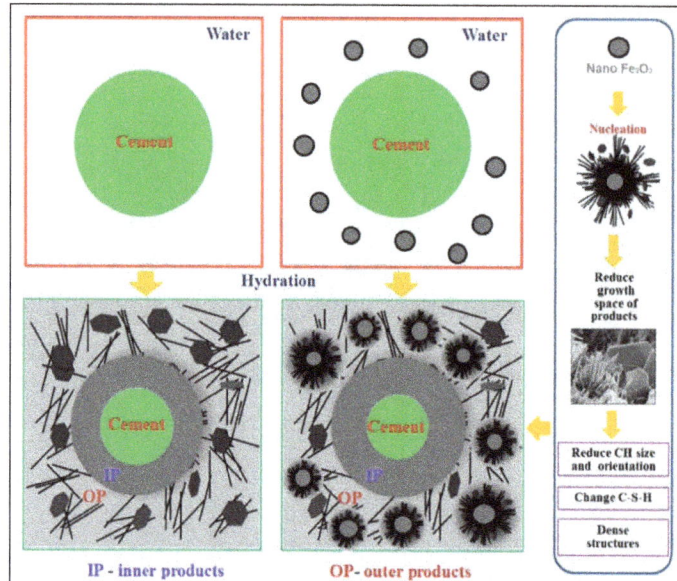

Models of influence of nano-Fe_3O_4 on the hydration products' growth.

Nanoindentation is a method increasingly used in the investigations of the structure of the cement composites matrices on the nanoscale. Most often used is the grid indentation method or statistical nanoindentation. The main binding phase is hydrated calcium silicate C–S–H with a heterogeneous structure. The C–S–H phase consists of 4–5 nm elementary spheres, appearing in the form of a colloid with different packing densities. Three main types (phases) of C–S–H are distinguished according to their densities: low-density (LD) C–S–H, high-density (HD) C–S–H, and the calcium hydroxide (CH)/C–S–H phase nanocomposite. The strong differentiation of the hardness and Young's modulus was observed for the particular phases when the cement composites were tested using the nanoindentation method. These results, together with statistical methods of analysis, enable the modeling of the composite nanostructure the ranges of the Young's modulus for the particular phases of the hardened cement pastes can be determined as follows:

1) Porous phase of low stiffness: a modulus below 10 Gpa

2) LD C–S–H phase of low stiffness: 20 ± 5 Gpa

3) HD C–S–H phase of high stiffness: 30 ± 5 Gpa

4) CH/C–S–H phase: 40 ± 5 Gpa

Horszczaruk studied cement pastes containing a 5% admixture of nano-Fe_3O_4 with a solid nanosilica shell, produced by the method described in Cendrowski et al. The cement pastes were produced using the Portland cement with $w/c = 0.5$, without a superplasticizer. Figure presents the results of the nanoindentation modulus tests for the paste without the admixture (R) and with a MN admixture. Histograms of the determined Young's modulus values were created for every specimen. Then, the obtained curves were distributed into the peaks, determining the probability of the distribution of the particular phases based on the Young's modulus values attributed to the phases. Clearly visible is a growth in the porous phase volume in the MN sample, equal to 26.1%. Despite this increase, significant growth of the volume of the CH/C–S–H phase (25.6%) was noted compared to the reference sample (5.7%). Also, the mean values of the Young's moduli of the CH/C–S–H phase increased by almost 20% compared to the reference sample. The content of the used nanomaterial (5% of the binder mass) appeared to be significant, causing an excessive increase in the porosity of the composite. According to Cendrowski this phenomenon can be attributed to the large specific surface area of nanoparticles, generating higher water demand. This would confirm the investigations described in Khoshakhlagh et al. and Nazari et al. where the increase in porosity was explained by the lack of space for $Ca(OH)_2$ crystal growth, which, in turn, leads to higher shrinkage and creep in the cement matrix. Continuing the experiment described in Horszczaruk et al, Horszczaruk et al. reported that a 3% admixture of nano-Fe_3O_4/SiO_2 appeared to be the optimum content both in terms of the structure of the cement matrix and the final compressive strength of the composite.

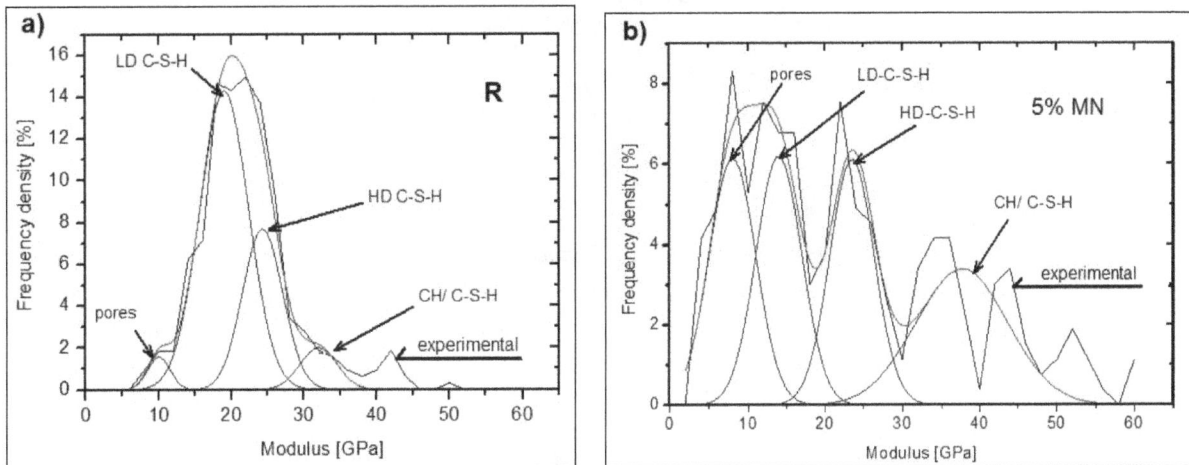

The experimental and theoretical probability distribution of the indentation modulus:
(a) reference sample (0% nano-Fe_3O_4/SiO_2) and (b) 5% of nano-Fe_3O_4/SiO_2.

The weakest link in cement composites, such as concrete or mortar, is the interfacial transition zone (ITZ), which occurs between the matrix and the aggregate or other filler. The ITZ significantly affects the macroproperties of the cement concretes, such as the strength, Young's modulus, or crack propagation. Strengthening of the transition zone is one of the reasons for the use of nanomaterials in cement composites. Such methods as SEM analysis, atomic force microscopy (AFM), or nanoindentation are employed for ITZ investigations. The nanoindentation method or AFM technique for evaluation of the ITZ between aggregate and paste, as well as between the particular phases of the cement matrix, require the extremely careful preparation of the tested surface of the composite. As described by Horszczaruk et al, the nanoindentation modulus and hardness of the cement matrix in the ITZ area of the aggregate-paste were measured. The tested objects were specially prepared specimens of the paste modified with 5% nano-Fe_3O_4 with a nano-SiO_2 shell on the border of the gravel aggregate with a diameter of 10–15 mm. For this aim, the gravel grain was placed on the bottom of a 20 mm cubic moulid and flooded with the paste. After 28 days of curing, the specimen was cut to half its height and carefully polished. The test results were compared with the reference sample made from the paste without MN admixture.

The hardness and indentation modulus were measured using the nanoindentation method, using the Nanoindenter XP produced by Agilent (Santa Rosa, CA, USA) and a three-sided pyramidal Berkovich indenter. At least 15 series, each with 8 imprints in the row, were performed for every specimen. The first imprint was placed about 10–20 μm from the edge of the aggregate. The distance between the imprints was about 20 μm. In figure the indenter's imprints inside the ITZ with the poorer hardness are marked in red. The imprints inside the zone with the higher hardness are marked in green. The results of testing the Young's modulus as a function of the distance from the aggregate grain surface are presented in figure.

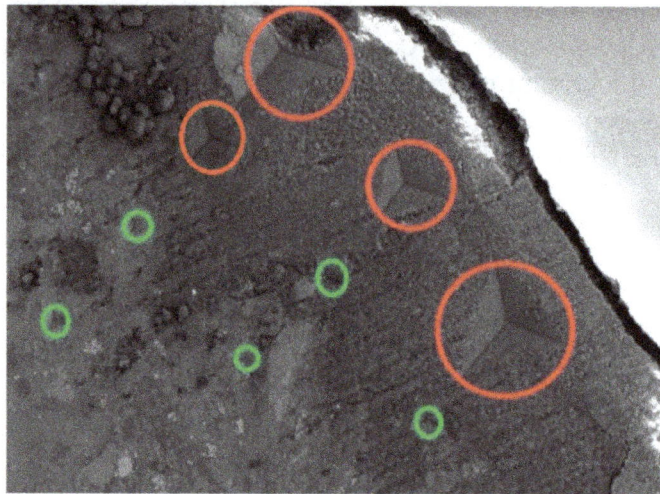

An SEM image of the imprints during the nanoindentation test.

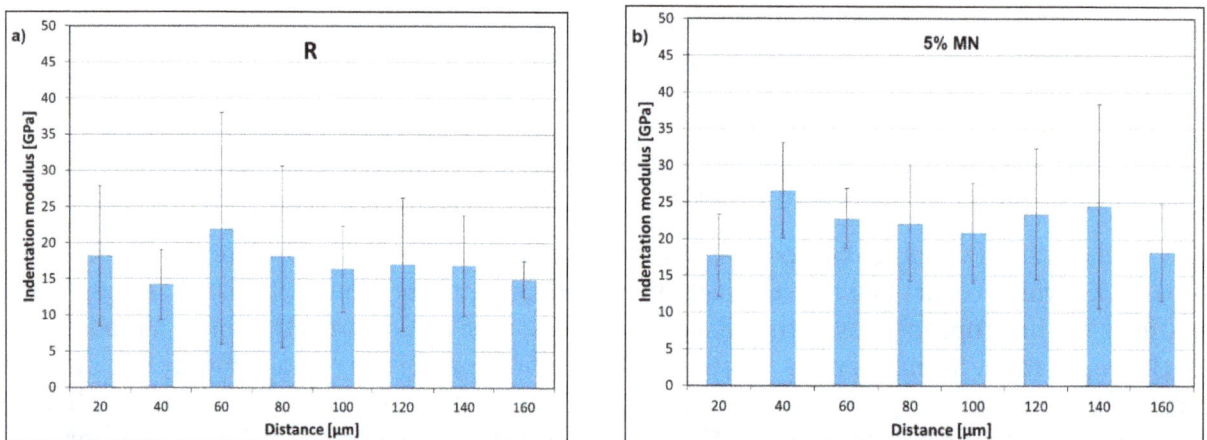

The results of measurements of the indentation modulus of the tested specimens in the transition zone: (a) reference sample (0% nano-Fe_3O_4/SiO_2) and (b) 5% nano-Fe_3O_4/SiO_2 content.

A 13% increase in the hardness due the contact zone aggregate-paste (ITZ) and by 29% inside the cement matrix, compared to the reference composite (R), was noted for the composite containing an admixture of nanoparticles (5% MN). The mean hardness and Young's modulus in the ITZ of the specimen containing an MN admixture were 0.59 and 17.72 MPa, respectively, and in the ITZ of the reference specimen (R), they were 0.52 and 14.27 MPa, respectively. The hardness and modulus outside the ITZ were 0.74 and 24.11 MPa, respectively, for the MN specimen, and 0.57 and 17.53, respectively, for the R specimen. A 22% increase in the value of the Young's modulus was noted for the MN specimen in the ITZ, and 37% inside the matrix, compared with the R specimen. The width of the ITZ in the specimen with 5% MN ranged from 25 to 40 μm. In the R specimen, the width of the ITZ was 40 to 70 μm. The improvement in the mechanical properties in the ITZ of the composite with NM can be explained by the use of a nano-SiO_2 shell, which—as demonstrated in numerous studies accelerates the hydration process, thus enhancing the strength and characteristics of the microstructure, which leads to an improvement in the adhesion in the aggregate-paste transition zone.

Mechanical Properties

Compressive Strength

Compressive strength is the most often investigated property of cement composites modified with MN. Only in a few papers, however, were the full test results provided, including the mean values and the standard deviations for the conducted measurements. Many authors only provide the relative strength, often presenting the data solely in graphical form, complicating comparison with the other research. The results of the compressive strength testing of the cement composites modified with MN are listed in table.

Table: The effect of MN on the compressive strength of cement composites.

Matrix Type/Type of Nanoparticle	Enhancement						Content of MN (wt %)
	After 3 Days		After 7 Days		After 28 Days		
	Abs. (MPa)	Rel. (%)	Abs. (MPa)	Rel. (%)	Abs. (MPa)	Rel. (%)	
paste/Fe_3O_4	43	0.00	55	3.77	74	4.22	0.05
	45	4.65	57	7.54	85	15.71	0.10
	48	11.62	66	24.52	67	−5.63	0.20
paste/Fe_3O_4/SiO_2	43	0.00	53	0.00	73	2.81	0.05
	43	0.00	56	5.66	78	9.85	0.10
	45	4.65	60	13.20	81	14.08	0.20
paste/Fe_3O_4	—	—	—	—	60	50.00	10.0
paste/Fe_3O_4	—	35.80	—	15.00	—	—	5.0 (fluid) 5.0 (powder)
	—	24.50	—	7.30	—	—	
mortar/Fe_3O_4	—	—	—	—	62.8	20.07	3.0
	—	—	—	—	54.3	4.59	5.0
mortar/Fe_3O_4/SiO_2	—	—	—	—	51.5	−1.53	3.0
	—	—	—	—	55.5	6.12	5.0
concrete/Fe_3O_4	—	—	—	—	119	28.93	1.5
concrete/Fe_3O_4	—	—	—	—	64	4.92	1.5
concrete/Fe_3O_4	—	—	28.0	19.15	36.4	6.43	1.0

Amin et al. investigated the compressive strength of cement composites modified with nano-Fe_3O_4 at 3, 7, 14, 28, and 90 days. The pastes were prepared using Portland cement (PC) as well as cement with a high content of blast-furnace slag (75% of the slag). The pastes were modified by adding nano-Fe_3O_4 in the amounts of 0%, 0.05%, 0.1%, and 0.3% of the binder mass. The water-binder ratio w/b coefficient was constant at 0.30. The results showed a fast rate of hydration during the period up to 14 days with the addition of nano-Fe_3O_4 in PC pastes. An increase in the compressive strength of the specimens, proportional to the nano-Fe_3O_4 content in the paste, was observed within this period. From 28 to 90 days of curing, the specimens containing 0.05% nano-Fe_3O_4 showed clear growth in the compressive strength compared to the reference PC sample. Only very slight increases in the strength, lower than that of the reference sample PC, were noted for the pastes with 0.1% and 0.3% nano-Fe_3O_4. According to Amin, the increase in the compressive strength in the presence of nano-Fe_3O_4 lower than that of the reference PC sample, especially during early hydration, could be attributed to the acceleration of the hydration reaction by nano-Fe_3O_4. In addition, the interaction of nano-Fe_3O_4 with the released free $Ca(OH)_2$ led to the formation of a hydrated product with a structure similar to that of Al-ettringite, designated Fe-ettringite, which had reasonable hydraulic characteristics. The Fe_3O_4 nanoparticles play the role of accelerators in the hydration process, which lead to an improvement in the microstructure of the modified composites.

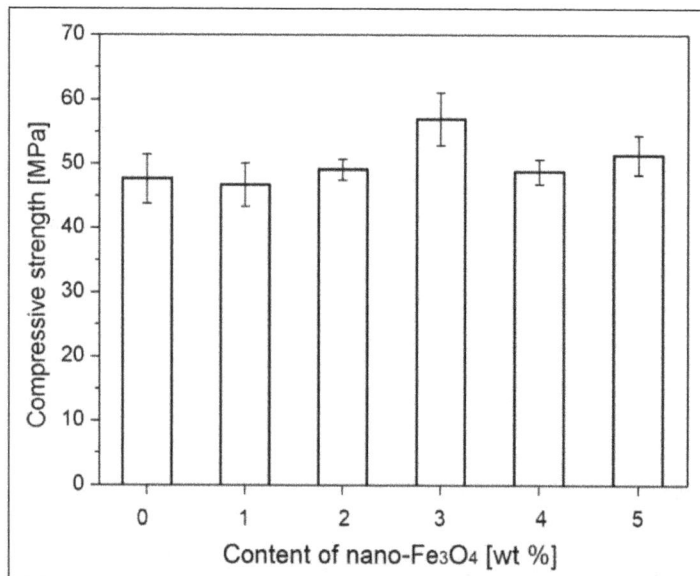

The compressive strength of cement mortars containing nano-Fe_3O_4 after 28 days of curing.

Bolhassani and Sayyahmanesh investigated cement pastes with similar composition, modified with nano-Fe_3O_4 and nano-Fe_3O_4/SiO_2 in the amounts of 0%, 0.05%, 0.1%, and 0.2% of the binder mass, with a w/b coefficient of 0.28. They determined the compressive strength of the pastes after 3, 7, and 28 days of curing. For the pastes containing small amounts of MN (0.05%), no increase in the compressive strength was observed within the first 7 days. However, significant growth in the compressive strength within the first 7 days of specimen curing was observed for the pastes containing 0.1% and 0.2% MN. Much higher increases in the compressive strength after 7 days than after 3 days of curing were observed for all tested specimens containing MN. The greater increases in the compressive strength compared to the reference paste were noted for pastes modified with an admixture of nano-Fe_3O_4 than in pastes modified with nano-Fe_3O_4/SiO_2, regardless of the age of the tested specimens. After 28 days of curing, all specimens with a MN admixture had higher compressive strengths than the reference paste, except for specimens with 0.2% nano-Fe_3O_4, which had significantly lower strength. This was caused by the poor mixing of the components in this series of specimens. Bolhassani, similar to Amin et al, attributed the increase in compressive strength to the role of magnetite nanoparticles in cores, on which the hydration products can adsorb, which, in turn, causes the compaction of the structure.

The tests of the compressive strength of the mortars modified with various amounts of nano-Fe_3O_4 were carried out by Sikora. They investigated the mortars with $w/c = 0.5$ containing 0%, 1%, 2%, 3%, 4%, and 5% nano-Fe_3O_4 in relation to the cement mass. Figure shows that a small amount of nano-Fe_3O_4 does not significantly influence the compressive strength of the cement mortars. However, with the increase in the nanoaddition to 2 or 3 wt %, a positive effect on the compressive strength was detected. The highest compressive strength was observed for the samples containing 3 wt % nano-Fe_3O_4. With increasing nano-Fe_3O_4 to 4 or 5 wt %, however, a reduction in the strength was noticed. Therefore, the sample containing 3 wt % nano-Fe_3O_4 (N3) is best. These results were confirmed by the previous findings of Yazdi and Amin, which showed that there is a certain amount of Fe_2O_3 or Fe_3O_4 that is beneficial for the cement composites. Exceeding this amount may result in the lowering of cement composite strength.

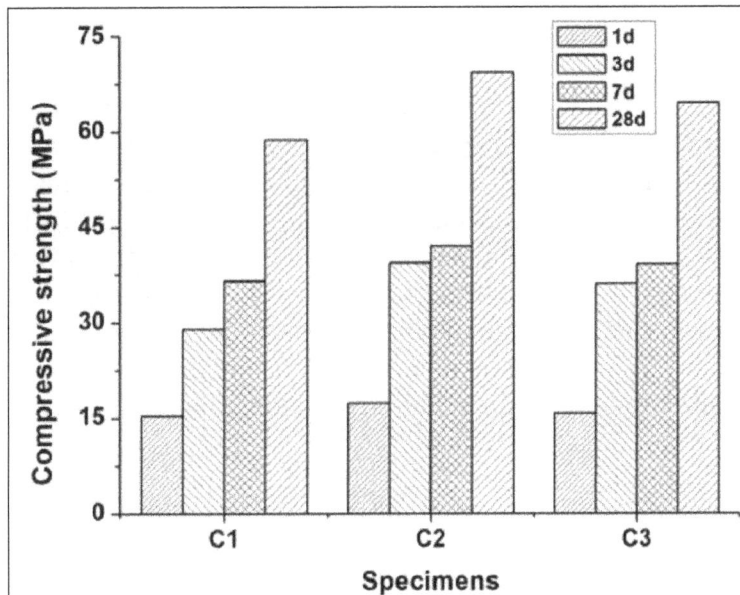

Compressive strength of the pastes: C1, reference paste; C2, paste containing nano-Fe_3O_4 in magnetic fluid form; and C3, paste containing nano-Fe_3O_4 in powder form.

Investigations into the influence of the nano-Fe_3O_4 application method on the compressive strength of the cement paste were performed by He et al. They prepared cement paste specimens with $w/c = 0.4$. The C1 specimen was a reference paste, C2 contained a 5% admixture of nano-Fe_3O_4 in the form of a magnetic fluid, and C3 contained a 5% admixture of nano-Fe_3O_4 in the form of a powder. Figure 20 shows the compressive strength of the specimens. The addition of nano-Fe_3O_4 both as a magnetic fluid and a powder increased the compressive strength of the cement paste. The effect was more obvious for nano-Fe_3O_4 in the form of the magnetic fluid and during the early stages. Compared to the control specimen C1, the strength of C2 increased by 13.4%, 35.8%, and 15.0% after 1, 3, and 7 days, respectively. Similarly, the strength of C3 increased by 2.7%, 24.5%, and 7.3% after 1, 3, and 7 days, respectively. He explained the increase in the compressive strength in the C2 and C3 by the fact that the nanoparticles provide very large surface areas and act as nucleation sites. The same improvement effect of nano-Fe_3O_4 on the strength was described by Shekari the dispersion of nano-Fe_3O_4 in the form of a magnetic fluid in the cement paste is better than that of nano-Fe_3O_4 in the powder form; thus, its effect on the mechanical properties is more obvious.

Flexural/Tensile Strength

Flexural strength and tensile strength are used to represent the toughness of the cement composite, which influence its brittleness. Sikora tested the flexural strength of the mortars containing 1–5% nano-Fe_3O_4. They found that the admixture of nano-Fe_3O_4 slightly worsened the tensile strength of the mortars ($w/c = 0.5$); however, this finding could be the result of the uneven distribution of the nanoparticles in the composite. The mortars containing 3% and 5% nano-Fe_3O_4 (RF3 and RF5, respectively) and nano-Fe_3O_4/SiO_2 (RFNS3 and RFNS5, respectively) were tested. The results were compared to the reference mortars containing no nanoparticles (R). The flexural strength was determined after 28 days of curing of the specimens. Similar effects in the improvement of the strength were observed for both types of the magnetite nanoparticles. Compared to the reference sample R, the flexural strength of RF3 increased by 8.5%, RF5 by 9.76%, RFNS3 by 8.5%, and RFNS5 by 12.19%.

Shekari and Razzaghi investigated the tensile strength of high-strength concrete (HSC) containing a 1.5% admixture of nano-Fe_3O_4. They noted a 26.3% increase in the 28-day tensile strength compared to the reference concrete. Jaishankar and Mohan obtained similar results from tensile strength testing for ordinary concrete with the same content of nano-Fe_3O_4 (1.5%). They observed an increase in the tensile strength of the specimens containing nano-Fe_3O_4 by about 25% compared to the reference concrete. In both investigations, the indirect method was used to determine the tensile strength of the concrete. With the limited number of studies, it is difficult to formulate specific conclusions. It can be seen, however, that the strengthening of the composite in the range of its flexural and tensile strength can be achieved, provided that the MN nanoparticles are properly distributed in the cement matrix.

Functional Properties

Electromagnetic Wave Absorption

Electromagnetic waves (EMW) have found a many uses in industrial production, wireless communication, military technology, and everyday life. However, electromagnetic radiation causes environmental contamination and is potentially harmful to health; it can also be a source of noise in the transmission of information. The negative effects of electromagnetic radiation have been a big concern, and developing electromagnetic-wave-absorbing materials is important in military and civil applications, such as stealth, microwave interference protection, and microwave darkrooms. Some researchers tried to add different absorbers to cement to improve EMW absorption by the cementitious building materials. The content of the absorber in the cement, necessary for efficient absorption, ranged from 10% to 30% of the mass, which significantly worsened the workability of the fresh composite and its mechanical properties. Moreover, the absorbers in powder form, used in the construction, often show a tendency of agglomerating, which further complicates their proper application in the composite.

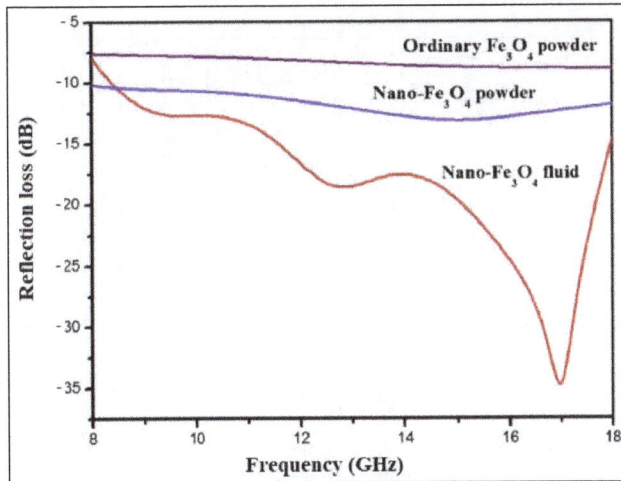

Reflection loss versus frequency for the cement composite containing various amounts of nano-Fe_3O_4 magnetic fluid.

The possibility of using nano-Fe_3O_4 in liquid form as the admixture to enhance EMW absorption (EMWA). The nano-Fe_3O_4 magnetic fluid was prepared using the co-precipitation method. The obtained liquid admixture containing nano-Fe_3O_4 was used to prepare the cement mortars. The mortars with $w/c = 0.4$ and nano-Fe_3O_4 content equal to 3%, 5%, and 7% of the cement mass were produced. The reflection loss (RL) was determined for the characterization of the EMWA of the hardened mortars. This is an important parameter for the evaluation of EMW absorption by the materials. The more negative the RL value, the higher the MWA absorption. The RL value is affected by many factors, such as the magnetic parameters and specific surface area of the absorber, structure and thickness of the cement composite, material, and EMW frequency.

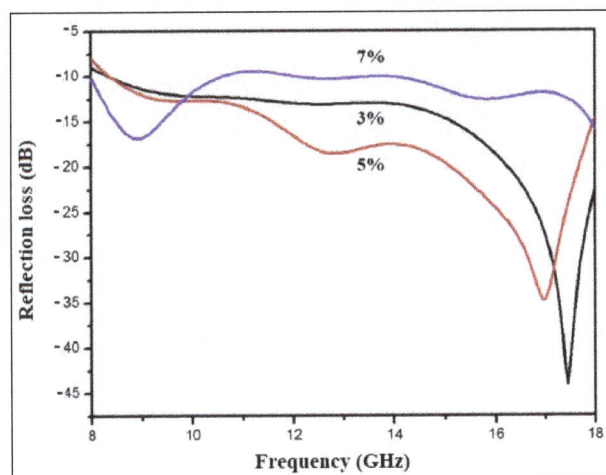

Reflection loss versus frequency for a cement composite containing 5% nano-Fe_3O_4 magnetic fluid, nano-Fe_3O_4 powder, and bulk Fe_3O_4 powder.

In figure shows the results of RL measurements of a cement composite with 5% nano-Fe_3O_4 magnetic fluid, nano-Fe_3O_4 powder, and bulk Fe_3O_4 powder. Compared to nano-Fe_3O_4 powder and bulk

Fe_3O_4 powder, the nano-Fe_3O_4 magnetic fluid significantly lowered the RL and broadened the absorption bandwidths due to its nanoscale particle size as well as its better dispersion in the cement paste. When the size of Fe_3O_4 is in the nanoscale range, its electronic polarization, ion polarization, and dipole polarization are enhanced.

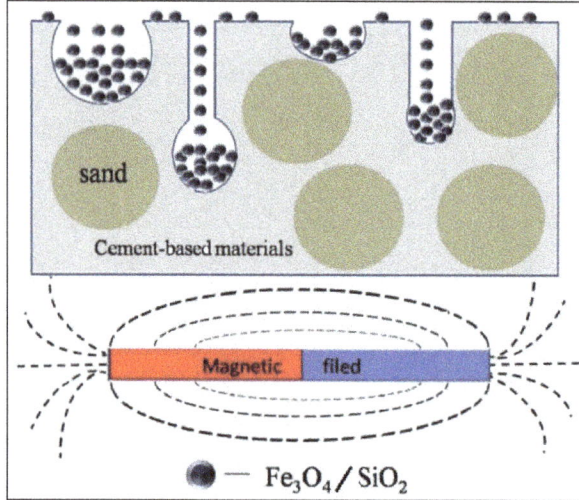

A schematic illustration of the surface treatment process of Fe_3O_4/SiO_2 nanoparticles under a magnetic field.

Similar results regarding the use of liquid nano-Fe_3O_4 as the admixture improving the EMWA of cement composites were reported by Wang. They used Fe_3O_4 nanoparticles and nanoparticles consisting of Fe_3O_4 with a solid SiO_2 shell synthesized using Stöber's method. The Fe_3O_4 and Fe_3O_4/SiO_2 nanoparticles were dispersed in water under the same molar concentration. The mortar specimens after 28 days of curing were dried to a constant mass and then soaked using the water dispersion of the nanomaterial. The specimens' surfaces were treated three times for 20 min. After soaking, the specimens were exposed to a magnetic field. Figure provides a schematic illustration of the surface treatment process of Fe_3O_4/SiO_2 particles under a magnetic field. Two hours later, the surface-treated cementitious materials were washed with water in order to remove the unbonded Fe_3O_4 and Fe_3O_4/SiO_2 nanoparticles from the surface.

The tests of the reflection confirmed that the use of the superficial treatment with nano-Fe_3O_4 and Fe_3O_4/SiO_2 improved the EMWA of the tested mortars. Better results were obtained for the mortars treated with nano-Fe_3O_4/SiO_2.

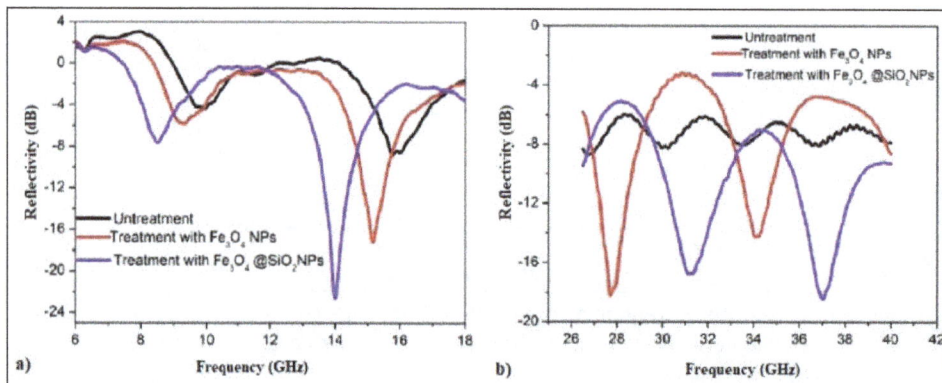

The reflectivity of cement mortar treated with Fe_3O_4 and Fe_3O_4/SiO_2 nanoparticles (NPs) at (a) 6–18 GHz and (b) 26–40 GHz.

Gamma-ray Shielding

A growing trend has been observed in the number of studies focused on searching for building materials that efficiently protect against gamma radiation. The basis of these studies is shielding concrete, manufactured using natural and artificial heavy aggregates. Other additions to concrete that could improve its shielding properties, including waste materials, such as silica fume, barite powder, magnetite powder, fly ash, or granulated ferrous waste, have also been sought. In this context, interest has grown in the various nanomaterials. The nanomaterials most often used for the modification of the cement composites are SiO_2, Fe_3O_4, and PbO_2. A resistance to high temperature is required for structural concretes exploited in nuclear power plants as well as in other objects possibly exposed to the radiation. The internal shielding layer of the concrete is often subject to heat from the reactor. For cooling water leakage from the reactor, the temperature can reach up to 360 °C. Fire only occurs in emergency or exceptional situations; therefore, the durability of structural concrete under high temperatures is very important, regardless of the object destiny. Concrete is more resistant to fire than other construction materials; however, increases in temperature above 200 °C negatively affect the mechanical properties of concrete. High temperatures also negatively influence the shielding properties of cement concrete.

One of the main parameters enabling the evaluation of the absorption of ionizing radiation is the linear attenuation coefficient μ, which is the relative diminishing of intensity of a radiation beam on the unit. In the investigations into the material's ability to absorb gamma radiation, the parameters HVL and TVL are also used. The HVL and TVL values represent the thickness of an absorber (x) that will reduce the gamma radiation to one-half and to one-tenth of its original intensity, respectively:

$$HVL = x1/2 = 1N2\mu$$

$$TVL = x1/10 = 1N10\mu$$

where μ is the linear attenuation coefficient (cm^{-1}).

The results of tests performed by Horszczaruk for cement pastes containing 5% and 10% nano-Fe_3O_4 in relation to the cement mass ($w/c = 0.5$) as well as the results of testing the cement concretes ($w/c = 0.35$) containing 3% nano-Fe_3O_4/SiO_2. The nanomagnetite with a grain diameter of 50–100 nm produced by Sigma Aldrich (637106, Darmstadt, Germany) was used in the tests. The concrete specimens were made from the Portland cement CEM I 42.5 R and natural pebble aggregate up to 16 mm, with the use of the polycarboxylic superplasticizer. The pastes and concretes specimens, after 28 days of curing, were exposed to heat in a medium-temperature oven at temperatures of 300 and 450 °C. The linear attenuation coefficient μ and the values of HVL and TVL were determined for pastes specimens (RP, reference paste; P5, paste with 5% MN; and P10, paste with 10% MN) and concrete specimens (RC, reference concrete and NC, concrete with 3% MN) after heating by exposing the specimens to gamma radiation. The specimens were irradiated by a gamma ray source of ^{137}Cs with an activity of 10 mCi and a photon energy of 0.662 MeV.

Table: The results of testing the shielding properties of cement pastes and concretes.

Symbol of Specimen	MN Content (wt %)	μ (cm⁻¹)			HVL (cm)			TVL (cm)		
		Temperature (°C)								
		20	300	450	20	300	450	20	300	450
RP	0	0.133	0.116	0.113	5.21	5.98	6.13	17.31	19.85	20.38
P5	5	0.134	0.118	0.114	5.17	5.87	6.08	17.18	19.51	20.20
P10	10	0.137	0.121	0.116	5.06	5.73	5.97	16.80	19.03	19.85
RC	0	0.186	0.179	0.179	3.73	3.85	3.87	12.38	12.79	12.86
NC	3	0.187	0.184	0.181	3.71	3.78	3.82	12.31	12.54	12.69

The μ values for the tested pastes and concretes are presented in figure. The higher the μ, the better the shielding properties of the material. In the temperature range of 20–450 °C, both pastes and concretes with an admixture of magnetite nanoparticles showed better shielding properties than the reference composites. This was also confirmed by the results of calculation of HVL and TVL, the values of which are lower for the concretes and pastes containing a MN admixture. As the total content of MN in the mass of the concrete is lower than 1%, the possibility of using of this type of nanoparticles in cementitious composites, particularly in repair materials, appears promising.

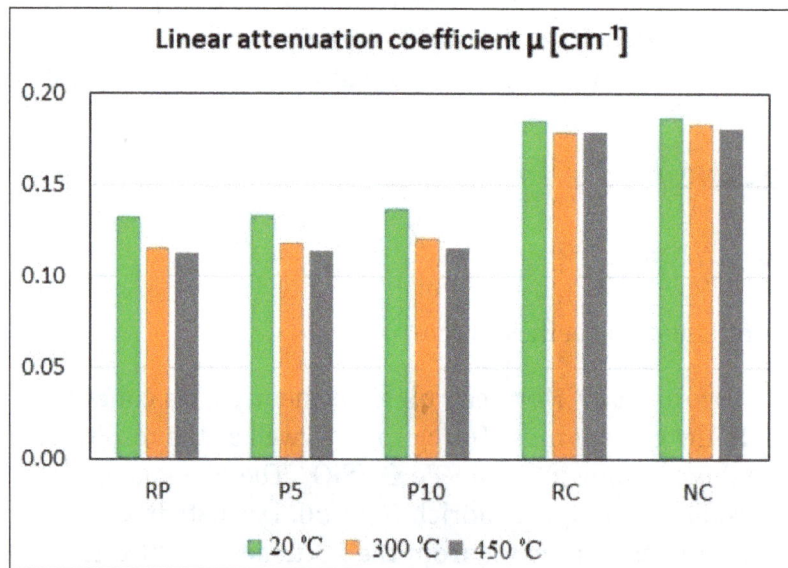

Linear attenuation coefficient μ for the tested cement pastes and concretes before and after exposure to 300 °C and 450 °C.

The value of μ decreases with increasing temperature; however, this decrease is small considering the relative change in the compressive strength as a function of specimen heating temperature. The influence of the modification of the concrete with MN particles is significant within the temperature range of 300–450 °C. At 600 °C, the destructive processes in the cement matrix begin, causing numerous cracks in the concrete and debonding of the matrix from the aggregate. This leads to a 40% decrease in the initial strength of the tested concretes.

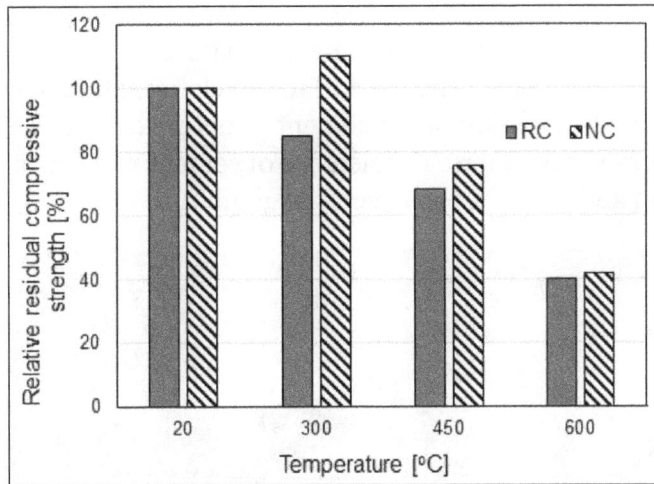

Relative changes in the compressive strength of tested concretes as a function of heating temperature (RC–reference concrete, NC–concrete with 3% of nano-Fe_3O_4/SiO_2).

Thermal Resistance of Cementitious Composites

The nanoparticles of iron oxides were tested with respect to the improvement in the thermal resistance of cement composites concerning the influence of Fe_2O_3 nanoparticles on thermal resistance, demonstrated that even small amounts (1% of the mass) of the nanomaterial can improve the fire resistance of cement pastes due to an increase in the residual compressive strength, lower mass losses, and compaction of the composite's microstructure. The presence of nano-Fe_2O_3 in cement matrices has been observed to cause a decrease in the length of the formed cracks. The influence of a 5% admixture of nano-Fe_3O_4 on the thermal resistance of cement pastes was analysed by Mijowska within a temperature range of 200–800 °C. At 450 °C, the specimens containing MN were demonstrated to be about 97–100% as strong as initially, whereas the unmodified specimens were only 80% as strong. At 600 °C, the strength of the specimens containing MN and the unmodified specimens were 59–64% of the initial strength, and at 800 °C, they were 25–30% as strong. The cracking of the surfaces of the pastes modified with MN, heated at 450 and 600 °C, was reduced compared to the unmodified pastes.

Nano-Fe_3O_4 is oxidized at higher temperatures, thus losing some of its properties, such as the ability to shield EMWA. For this reason, various methods of improving nano-Fe_3O_4 stability were examined, for example, by synthesizing the nanosized protective layers on the surface of particles.

Incorporation of molecular hybrids into cement-based composites allows for the combination of the positive influences of several nanomaterials due to the synergic effect, leading to the simultaneous improvement in various properties. This effect was achieved by Bolhassani by incorporating the core-shell nanostructures into the cement composite. The magnetite particles acted as cores and were covered by the silica shell prepared with the Stöber's method. This type of nanocomposite structure did not cause a significant worsening in the workability, in contrast to nanomaterials introduced separately. The use of a magnetite-silica nanocomposite caused a significant increase in the relative compressive strength compared to the unmodified material.

The results of thermal resistance tests of the cement mortars modified with nano-Fe_3O_4/SiO_2. The cement mortars contained 3% and 5% nano-Fe_3O_4/SiO_2. The specimens were heated to 200–800

°C. The obtained results were compared with those obtained for the specimens modified with the same amounts of nano-Fe_3O_4 and without the nanoadmixture. Fe_3O_4/SiO_2 nanoparticles improve the thermal resistance more by decreasing the surface crack formation in cement mortars (except at 800 °C, where the total degradation of the cement structure occurred) than the pristine Fe_3O_4 nanoparticles. Fe_3O_4 improves the thermal resistance of cement mortars at 450 °C and 600 °C. At lower temperatures (200 and 300 °C), the effect is negligible.

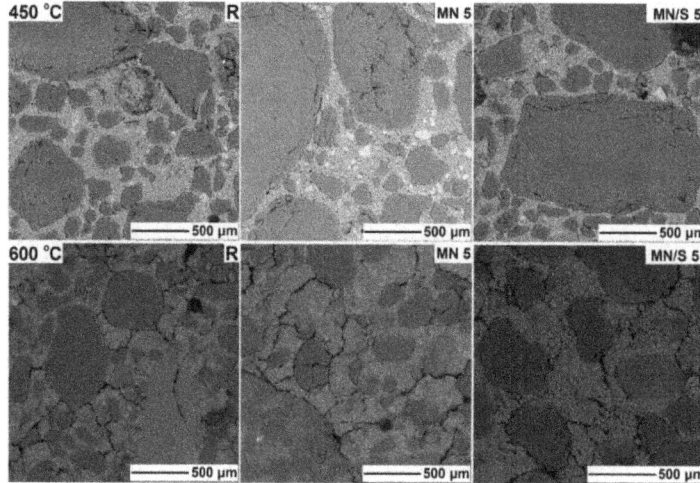

SEM micrographs of cement mortars containing 5% nano-Fe_3O_4 (MN 5), 5% nano-Fe_3O_4/SiO_2 (MN/S 5), and reference specimen (R) exposed to 450 °C (top) and 600 °C (bottom).

The thin silica shell is capable of extending the temperature range by positively influencing the effect of the iron oxide nanoparticles on the cement mortars' residual compressive strength. The pristine Fe_3O_4 nanoparticles significantly improve the residual compressive strength only at 450 °C. The cement mortars containing Fe_3O_4/SiO_2 exhibited improved residual compressive strength within the temperature range of 200–600 °C.

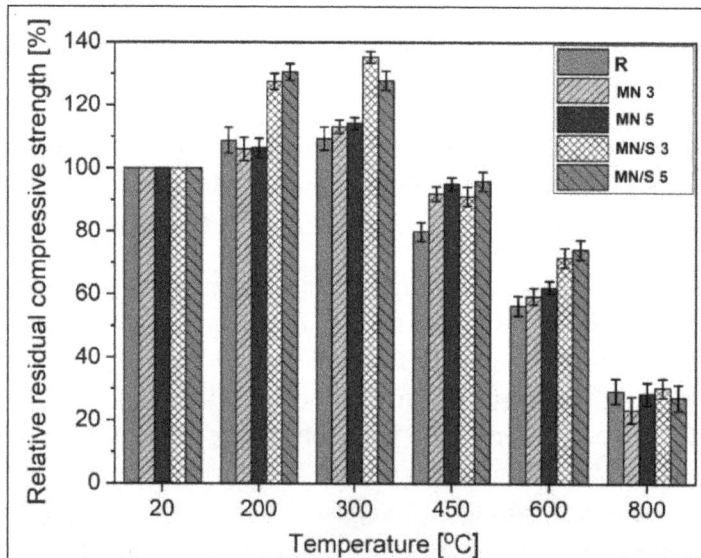

Relative residual compressive strengths of cement mortars as a function of temperature (R, reference; MN 3 and MN 5, cement mortars containing 3% and 5% nano-Fe_3O_4, respectively; MN/S 3 and MN/S 5, cement mortars containing 3% and 5% nano-Fe_3O_4/SiO_2, respectively).

References

- Ceramic-Matrix-Composites: researchgate.net, Retrieved 14 June, 2019

- Fabrication-of-ceramic-matrix, composites-by-polymer-infiltration-and-pyrolysis: substech.com, Retrieved 20 February, 2019

- Composites-by-chemical-vapor-infiltration, composites-by-polymer-infiltration-and-pyrolysis: substech.com, Retrieved 23 July, 2019

- Composites-by-liquid-phase-infiltration, composites-by-polymer-infiltration-and-pyrolysis: substech.com, Retrieved 3 March, 2019

- Composites-by-Direct-Oxidation-Process, Fabrication-of-Ceramic, chemical-engineering: idc-online.com, Retrieved 8 August, 2019

- Composites-by-slurry-infiltration, fabrication-of-ceramic-matrix: substech.com, Retrieved 29 April, 2019

- Composites-by-sol-gel-process, fabrication-of-ceramic-matrix: substech.com, Retrieved 19 January, 2019

4

Carbon-Carbon Composites

The composite material which is composed of a matrix of graphite reinforced by carbon fiber is called carbon-carbon composite. Its processing is done using either gas phase impregnation or liquid phase impregnation process. The topics elaborated in this chapter will help in gaining a better perspective about carbon-carbon composite as well as its processing.

Carbon-carbon composites (CC Composites) are a new class of engineering materials that are ceramic in nature but exhibit brittle to pseudoplastic behaviotir. Carbon-carbon is a unique all-carbon composite with carbon fibre embeded in carbon matrix and is known as an inverse composite. Due to their excellent thermo-structural properties, carbon-carbon composites are used in specialised application like re-entry nose-tips, leading edges, rocket nozzles, and aircraft brake discs apart from several industrial and biomedical applications.

As in all composites, the aim is to combine the advantage of high specific strength and stiffness of carbon fibres with the refractory properties of carbon matrix. When the fibres are laid in near-net shapes with multidirectional reinforcements, the result is an ideal high temperature structure.

Unique Features

These composites are the best among all high temperature materials because they are thermally stable and do not melt up to 3000 °C, have high thermal conductivity and low thermal expansion (thus having high resistance to thermal shock) and retain their mechanical strength to the end. Also, these composites maintain good frictional properties over the entire temperature range with low wear. They have high fracture toughness and do not fracture in a brittle manner like conventional ceramics. A multimode mechanism of fracture occurs where the fibres break as well debond. The brake discs for high speed aircrafts like Mirage 2000, Concorde, Airbus-320 are Some of the example where the favourable frictional properties were put to use. The first generation CC composites had the limitation of proneness to oxidation over long exposures. However, with the advent of second generation oxidation-resistant composites, this limitation was overcome. These unique features made it the most favourite material for re-entry nosetips, leading edge material for space shuttle wings, rocket nozzles, thrust vectoring nozzles using CC ball and socket joints and high performance turbojet engines.

International Status of CC Composites

CC composites have their origin in the jet vanes used in the German V_2 rockets. The jet vanes were made of graphite which eroded rapidly and had limited life. Subsequently, pyrolitic graphite, A TJ graphites were used which when reinforced with carbon fibres gave birth to CC composites. Major work is going on in USA, France and former Russia. There is a great demand for CC brake discs for aircrafts, trains, trucks and even racing cars. The pioneers in this field are Bendix in USA and Dunlop in UK. Nigrafitc, Moscow, is the leading organisation in former USSR. Little is known about its products, while Germany and Japan are in the race for industrial applications. Israel, Taiwan and Egypt were reported to have initiated some R&D activities. A study conducted by Dr Robert A; Meyer on CC composite research in the Far East indicates that, apart from India, as many as 18 institutions in Japan, 4 in China, 3 in Taiwan, 3 in Korea, 2 in Australia are carrying out active research on different aspects of CC technology. His assessment indicates that the collective research efforts in Japan and other Asian countries will improve and, in time. Surpass the research activities in the United States if the present financial support trend continues. Internationally, the stress is on industrial applications. CC composites are a candidate material in glass industry, furnace and semiconductor industry as well as for preventing corrosion in chemical plants. One interesting and innovative use is as tooling material for metallurgical superplastic stage forging process. Other high tech applications are as heat exchanger tubes for helium cooled high-temperature 'Nuclear reactors, high temperature crucibles, fastners, load bearing plates, rods and heating elements. Elemental carbon is "known to have the best biocompatibility with blood and soft tissues. Thus it finds use in hip bone endoprosthesis, bone plates, osteosynthesis and artificial heart valves.

Carbon Fiber

Carbon fibers are novel high-performance materials. They could be described as fibers containing at least 90 % carbon obtained by the controlled pyrolysis of appropriate fibers. Edison in 1879 found that carbon fibers can be used as carbon filaments in electric lamps. Since the early work of Edison, the carbon fibers have been investigated and used intensively because they generally have excellent tensile properties, low densities, high thermal and chemical stabilities in the absence of oxidizing agents, good thermal and electrical conductivities, and excellent creep resistance.

In recent years, the carbon fiber industry has been growing steadily to meet the demands arising from different applications such as aerospace (aircraft and space systems), military, turbine blades, construction, lightweight cylinders and pressure vessels, medical, automobile, and sporting goods. Furthermore, in the case of the carbon fibers, the range of their applications would depend on the type of precursors used to produce the carbon fibers. Consequently, many types of precursors have been studied to produce the carbon fibers. The ideal features of precursors required to manufacture carbon fibers are easy conversion to carbon fiber, high carbon yield, and cost-effective processing. From this perspective, the following four types of precursors have been widely used, which have also proven to be the most popular ones.

1. Acrylic precursors: They have been successfully used for carbon fiber preparation by most industrial manufacturers for quite some time now. These acrylic precursors contain >85 % acrylonitrile (AN) monomer. In particular, polyacrylonitrile (PAN) is the most popular acrylic precursor, which is used widely to produce the carbon fibers.

2. Cellulosic precursors: They contain 44.4 % carbon. However, in practice, the reaction is more complicated than mere dehydration, and the carbon yield is only approximately 25–30 %.

3. Pitch-based precursors: They have a yield of 85 %, and the resultant carbon fibers from these precursors show a high modulus owing to the more graphitic nature. On the other hand, the pitch-based carbon fibers have poorer compression and transverse properties compared to the PAN-based carbon fibers.

4. Other forms of precursors: Vinylidene chloride and phenolic resins as precursors for the manufacture of carbon fibers have been investigated, but were not found to be commercially viable.

Acrylic Precursors

Various acrylic precursors have been utilized to produce carbon fibers for use in carbon-fiber-based composite applications owing to certain desirable physical properties. The acrylic precursors for the carbon fiber industry originated from the companies that were established commercial-scale producers of textile-grade acrylic fibers. This was because the carbon fibers were produced through the pyrolysis of the acrylic fibers. Therefore, the manufacturers of carbon fibers could most readily adapt the existing technology for precursors to manufacture carbon fibers. The manufacturing processes of acrylic precursors by various manufacturers are listed in table. In particular, the resultant carbon fibers from acrylic precursors such as PAN-based carbon fibers have been widely used as reinforcing materials in automobile, aerospace, recreational, and various other industries.

PAN Precursors

PAN-based polymer precursors for carbon fibers could be primarily classified into pure homopolymer and comonomers. Generally, the comonomers are widely used in PAN-based polymer precursors to manufacture carbon fibers.

Table: Manufacturing processes for acrylic precursors.

Manufacturers	Trade name	Type of polymerization	Solvent	Typical % of polymer
Accordis	Courtelle	Continuous solution	NaCNS	10–15
Asahi	Cashmilon	Continuous aqueous dispersion	H_2O/HNO_3	8–12
Hexcel	Exlan	Continuous aqueous dispersion	H_2O/ NaCNS	10–15
Mitsubishi	Finel	Continuous aqueous dispersion	H_2O/ DMAC	22–27
Toho	Beslon	Continuous solution	$ZnCl_2$	8–12
Toray	Toraylon	Batch solution	DMSO	20–25

The homopolymer PAN product is slightly difficult to process into carbon fibers because the initial oxidation stage of the process cannot be easily controlled owing to the sudden and rapid evolution of heat, coupled with a relatively high initiation temperature. Such heat could result in poor properties of carbon fibers owing to the chain scission from the thermal shock. It is known that the homopolymer PAN product has never been used as a precursor for manufacturing carbon fibers. To overcome the resultant poor properties in the carbon fibers due to the rapid evolution of heat, the exothermic reaction should be adequately controlled using suitable comonomers such as itaconic acid.

On the other hand, as mentioned above, the comonomers could exert a significant effect on the stabilization process, thereby enhancing the segmental mobility of the polymer chains resulting in better orientation and mechanical properties of the precursor and manufactured carbon fibers. In addition, the selection of suitable comonomers could reduce the initial temperature of cyclization in the manufacturing process of carbon fibers. Therefore, several researchers have discussed the effect of comonomer composition on the properties of the PAN precursor and resultant carbon fibers.

Vinyl esters such as vinyl acetate (VAc), methacrylate (MA), and methyl methacrylate (MMA) could be used as comonomers for AN, though VAc might not be an appropriate carbon fiber precursor. These comonomers act like a plasticizer and break up the structure to make the polymer more readily soluble in the spinning solvent, thereby improving the quality of spinning, modifying the fiber morphology and where appropriate, and improving the rate of diffusion of the dye into the fibers. However, the comonomers should be minimally used for establishing good properties in the carbon fibers because they could affect the cyclization step in the manufacturing process of carbon fibers. Table lists the possible comonomers. These monomers have similar reactivities, and the resulting polymer compositions will have more or less the same composition as the monomers in the feed.

Carboxylic acids could be used as effective comonomers because their presence affects the ease of oxidation, exothermicity, and carbon yield of the precursor. In addition, itaconic and methacrylic acids have been confirmed to be most effective comonomers for reducing the exothermicity.

Table: Comonomers for PAN precursors.

Class	Comonomer	Structure
Acids	Acrylic acid	$CH_2=CHCOOH$
	Itaconic acid	$CH_2=C(COOH)CH_2COOH$
	Methacrylic acid	$CH_2=C(CH_3)COOH$
Vinyl esters	Methyl acrylate	$CH_2=CHCOOCH_3$
	Ethyl acrylate	$CH_2=CHCOOC_2H_5$
	Butyl acrylate	$CH_2=CHCOO(CH_2)_3CH_3$
	Methyl methacrylate	$CH_2=C(CH_3)COOCH_3$
	Ethyl methacrylate	$CH_2=C(CH_3)COOC_2H_5$
	Propyl methacrylate	$CH_2=C(CH_3)COO(CH_2)_2CH_3$
	Butyl methacrylate	$CH_2=C(CH_3)COO(CH_2)_3CH_3$
	β-Hydroxyethyl methacrylate	$CH_2=C(CH_3)COOCH_2CH_2OH$
	Dimethylaminoethyl methacrylate	$CH_2=C(CH_3)COOCH_2CH_2N(CH_3)_2$
	2-Ethylhexylacrylate	$CH_2=CHCOOCH_2(C_2H_5)(CH_2)CH_3$
	Isopropyl acetate	$CH_3COOC(CH_3)=CH_2$
	Vinyl acetate	$CH_3COOCH=CH_2$
	Vinyl propionate	$C_2H_5COOCH=CH_2$
Vinyl amides	Acrylamide	$CH_2=CHCONH_2$
	Diacetone acrylamide	$CH_2=CHONHC(CH_3)_2CH_2COCH_3$
	N-methylolacrylamide	$CH_2=CHONHCH_2OH$

Vinyl halide	Allyl chloride	$CH_2=CHCH_2Cl$
	Vinyl bromide	$CH_2=CHBr$
	Vinyl chloride	$CH_2=CHCl$
	Vinylidene chloride (1,1-dichloroethylene)	$CH_2=CHCl_2$
Ammonium salts of vinyl compounds	Quaternary ammonium salt of aminoethyl-2-methylpropenoate	$CH_2=CH(CH_3)COOC_2H_4NH_2$
Sodium salts of sulfonic acids	Sodium vinyl sulfonate	$CH_2=CHSOONa$
	Sodium p-styrene sulfonate (SSS)	$CH_2=CH\!-\!\bigcirc\!-\!SO_3Na$
	Sodium methallyl sulfonate (SMS)	$CH_2=C(CH_3)CH_2SO_3Na$
	Sodium-2-acrylamido-2-methyl propane sulfonate (SAMPS)	$CH_2=CHCONH(CH_2)_3SO_3Na$
Others	Methacrylonitrile	$CH_2=C(CH_3)CN$
	2(1-Hydroxylalkyl) acrylonitrile	$CH_2=\underset{\underset{R-CHOH}{\mid}}{C}-X$ (where $R = -CH_3$ or $-C_2H_5$ and $X = -CN$ or $-COOH$)
	Allyl alcohol	$CH_2=CHCH_2OH$
	Methallyl alcohol	$CH_2=C(CH_3)CH_2OH$
	1-Vinyl-2-pyrrolidone	
	4-Vinylpyridine	
	2-Methylene glutaronitrile	$NCCH_2CH_2C=CH_2CN$

The superiority of itaconic acid results from the presence of two carboxylic acid groups, which increases the possibility of interacting with a nitrile group. In other words, if one carboxylic acid group of itaconic acid moved away from an adjacent nitrile group owing to the dipole–dipole repulsion, the other carboxylic acid group could move into the vicinity of a nitrile group, thereby facilitating participation in the cyclization process.

Polymerization Methods for Production of PAN-based Precursors

Acrylic precursors such as AN and comonomers are initiated by free radical reaction and can be polymerized by one of the several methods listed below:

1. Solution polymerization: This method is used to prepare AN and comonomers by dissolving a monomer and catalyst in a nonreactive solvent. During this reaction, the solvent absorbs the heat generated by the chemical reaction and controls the reaction rate. The solvent used in the polymerization process usually remains as a solvent for the acrylic precursors (reactant). This process is only suitable for the creation of wet polymer types because the

excess solvent is difficult to remove. During this reaction, the solvent absorbs the heat generated by the chemical reaction and controls the reaction rate. The solvent used in the polymerization process usually remains as a solvent for the acrylic precursors (reactant). This process is only suitable for the creation of wet polymer types because the excess solvent is difficult to remove. Although the removal of excess solvent is possible using distillation, this method is usually not commercially viable. This polymerization method for the preparation of PAN-based precursors offers a few advantages as along with one major disadvantage.

2. Bulk polymerization: This method is the simplest and direct way of synthesizing polymers. This polymerization is carried out by adding a soluble initiator to a pure, liquid monomer. The initiator dissolves in the monomer, and the reaction is initiated by either heating or exposure to radiation. The mixture becomes viscous as the reaction proceeds. At that instant, the reaction is exothermic and a reactant with a broad molecular weight distribution is produced. Therefore, this polymerization method for the preparation of PAN-based precursors is used in the small-scale manufacturing process, wherein it is easy to remove the reaction heat.

3. Emulsion polymerization: This method is a type of radical polymerization, which is usually carried out with an emulsion containing water, monomer, and surfactant. The PAN-based precursors are often made commercially using emulsion polymerization.

4. Aqueous dispersion polymerization: This method is useful to prepare micro- or submicron-scale monodisperse polymer particles in a single step. In this polymerization method, all reaction materials are dissolved in the reaction medium in the initial stage of the polymerization. Then, insoluble spherical polymer particles, stabilized using steric molecules, are formed and dispersed in the reaction medium. PAN-based precursors could be prepared using this polymerization method with an ionic monomer to attain a narrow particle size distribution with a mean diameter as low as approximately 3 ± 1.5 μm.

Manufacture of Carbon Fibers from PAN-based Precursors

The PAN-based polymers are the optimum precursors for the carbon fibers owing to a combination of tensile and compressive properties as well as the carbon yield. The PAN-based fibers were first developed by Dupont in the 1940s for use in the textile fiber. The thermal stability of PAN-based fibers was an important factor in expanding the application of fibers. Later, this property led to further research on the heat treatment of PAN fibers. In the early 1960s, PAN fibers were first carbonized and graphitized by Shindo at the Government Industrial Research Institute, Osaka, Japan, and these, in the early 1960s, PAN fibers were first carbonized and graph 112 GPa, respectively. The process involved employing tension in both stabilization and carbonization steps. According to Toray, Shindo's patent was licensed to Toray in 1970 by the Japanese Ministry of International Trade and Industry (MITI) to produce PAN-based Torayca carbon fibers. During the 1960s, Watt and Johnson at the Royal Aircraft Establishment, England, and Bacon and Hoses at Union Carbide, the USA, also developed a method for producing carbon fibers from PAN.

The manufacturing steps for producing the carbon fibers from PAN could be categorized as follows as shown in figue. Polymerization of PAN-based precursors, spinning of fibers, thermal stabilization, carbonization, and graphitization. The manufacturing steps for producing the carbon fibers from PAN-based precursors could be categorized as follows as shown in figure. polymerization of PAN-based precursors, spinning of fibers, thermal stabilization (oxidation), carbonization, and graphitization figure. In addition, these PAN-based copolymers containing 2–15 % acrylic acid, methacrylic acid, MA, and/or itaconic acid are generally used as precursors of carbon fibers produced through carbonization because the use of comonomers affects the molecular alignment and stabilization conditions. The typical carbon yield from PAN-based precursors is approximately 50–60 %.

The typical manufacturing steps involved in the production of carbon fibers from PAN-based precursors are listed below:

1. Polymerization of PAN-based precursors and spinning of fibers: The PAN polymer precursor has been widely used as the basic backbone of chemical structure for spinning precursor fibers. Figure shows the chemical structure of PAN. In addition, the commercial PAN-based polymer precursors used for spinning the precursor fibers usually contain approximately 2–10 % of a comonomer such as methyl acrylate (MA), MMA, or itacon acids (ITA). Most carbon companies manufacture their own precursors by in-house technologies. Consequently, the composition of the PAN polymer precursor is not well known because it can control the properties of the final products.

Generally, the PAN polymer precursors contain polar nitrile groups and have a high melting point, resulting from strong intermolecular interactions. Therefore the PAN polymer precursors tend to degrade before the temperature reaches their melting point. The spinning of PAN fibers in the carbon fiber industry is performed using traditional manufacturing techniques of acrylic textile fibers. Wet spinning is used in most of the commercial manufacturing processes of carbon fibers produced from PAN-based polymer precursors. However, it is gradually being replaced by dry jet wet (air gap) spinning. The melt spinning of PAN-based polymer precursors has been previously practised; however, it has yet to become an acceptable manufacturing process of carbon fibers, commercially. Figure shows the typical layout of a plant for processing PAN-based fibers.

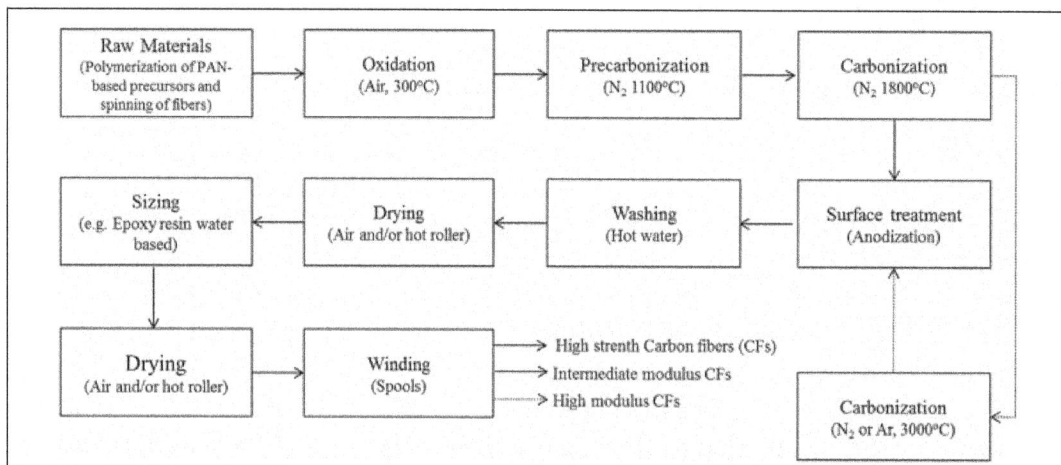

Schematic for manufacturing process of carbon fibers from PAN-based precursors.

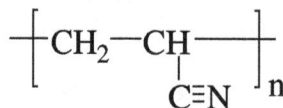

$$\left[CH_2{-}CH \atop \quad\quad C{\equiv}N \right]_n$$

Chemical structure of PAN.

2. Thermal stabilization (oxidation): This process is critical to obtaining high-quality carbon fibers and could take up to several hours, depending on the temperature, precursor diameter, and precursor fiber characteristics. Proper conditions such as heating rate, time, and temperature of heating should be established for the optimum stabilization of each precursor. The PAN-based polymer precursor is stabilized by controlled low-temperature heating over the range 200–300 °C in air to convert the precursor to a form, which could be further heat-treated without either the melting or fusion of the fibers.

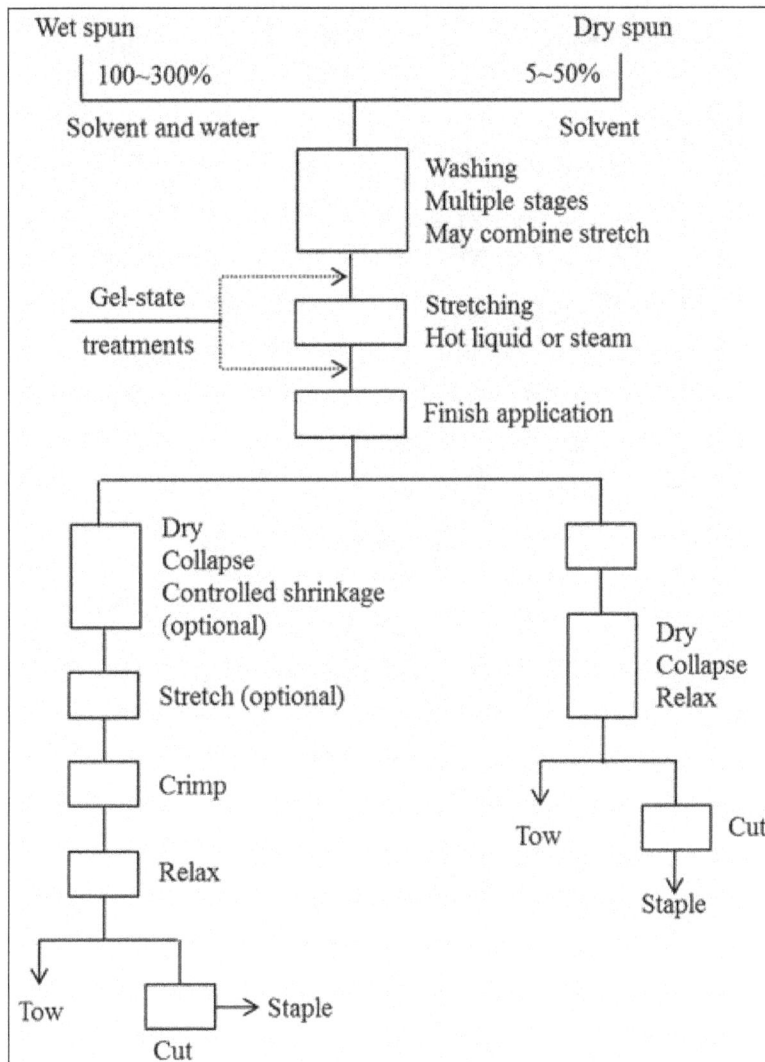

Typical layout of a plant for processing PAN-based fibers.

In this process, the linear molecules of PAN-based polymer precursor are first converted into cyclic structures. However, the cyclization is a complicated process, the mechanism of which is still unclear. In general, the most widely known reaction mechanism is shown in figure.

3. Carbonization and graphitization: The carbonization and graphitization of thermally stabilized fibers are carried out in an inert atmosphere containing gases such as nitrogen (N_2) or Ar. Generally, N_2 is the preferred gas, but Ar is used despite being eight times more expensive. This is because Ar provides improved strength to the carbon fiber owing to the higher density and viscosity

of argon. The temperature of carbonization is usually determined by the type of application of the resulting carbon fibers. For high-strength applications, the carbonization temperature over the range 1,500–1,600 °C is preferred because at temperatures above 1,600 °C, a decrease in the tensile strength occurs. On the other hand, an additional heat treatment above 1,600–1,800 °C and up to 3,000 °C, i.e., graphitization process, is required to obtain a high modulus in the carbon fibers Nitrogen cannot be used at temperatures above approximately 2,000 °C owing to its reaction with carbon to form cyanogen. The heating rate and retention time during carbonization are different depending on the type of the precursor and stabilization conditions. Figure shows the typical tensile strength (GPa) of the PAN-based carbon fibers depending on maximum carbonization temperature.

The carbonization and graphitization of the thermally stabilized fibers is a two-step process, i.e., low-temperature carbonization and high-temperature graphitization, depending on the requisite properties of the carbon fibers. Bromley et al. confirmed that the gases evolved during the carbonization of PAN-based carbon fibers over the low-temperature range 200–1,000 °C, and the observed gases are listed in table.

In figure shows the schematic of the graphite structure. Dehydrogenation joined the ladder molecules to form graphite-like ribbons; however, denitrogenation made the ribbons to form sheet-like structures. On the other hand, the high carbonization temperature caused the ordered structure to grow in both thickness and area, increased the crystalline orientation in the fiber direction and reduced the interlayer spacing and void content. In addition, the graphite structures could further grow at higher temperatures resulting from the elimination of N_2.

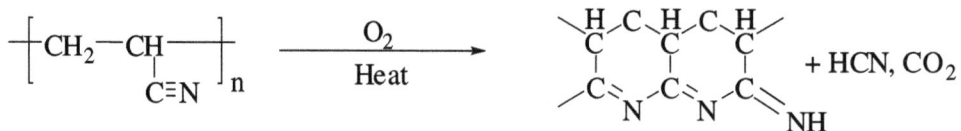

Mechanism of cyclization of PAN polymer precursor.

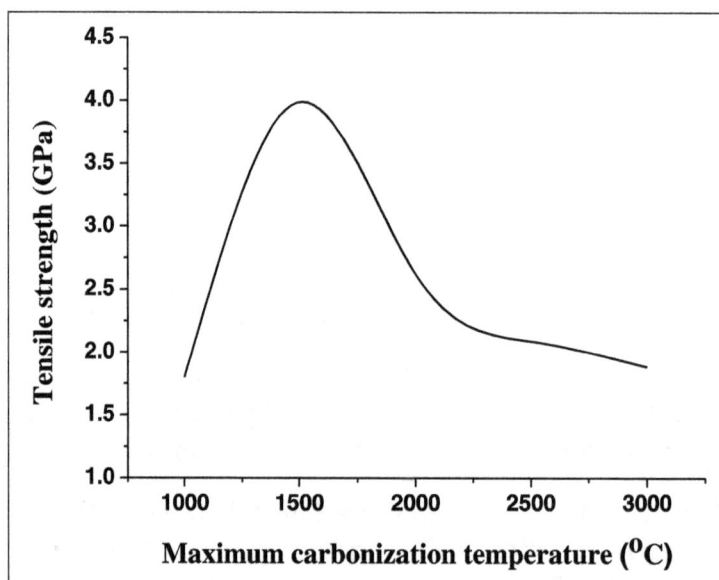

Typical tensile strength (GPa) of PAN-based carbon fibers depending on maximum carbonization temperature.

Table: Gases evolved during carbonization of PAN-based carbon fibers.

Temperature (°C)	Observation	Interpretation
220	HCN evolved and O_2 chemically bonded	Ladder polymer formation and oxidation of polymer
260	Little changed. No modulus increased	No chain scission
300	Large CO_2 and H_2O evolution; also CO, HCN, and some nitriles. No modulus increased	CO_2 from –COOH groups in oxidized polymer No cross-linking
400	CO_2, H_2O, CO, HCN, and NH_3 evolved. Small evolution of C3 hydrocarbons and nitriles Modulus increased	Cross-linking by intramolecular H_2O elimination
500	Increased H_2 evolution. Some NH_3 and HCN evolved. Modulus increased	Cross-linking by dehydrogenation
600	Reduced H_2 evolution. HCN and trace N_2	Cross-linking by dehydrogenation
700	N_2, HCN, and H_2 evolution. Modulus increased	Cross-linking by dehydrogenation and evolution of N_2
800	Large increase in N_2, H_2, and HCN still evolved. Modulus increased	Cross-linking by evolution of N_2
900	Maximum evolution of N_2, some H_2, and traces of HCN. Modulus increased	Cross-linking by N_2 elimination
1,000	N_2 evolution decreases to approximately the same level as that at 800 °C. Trace H_2 evolved. Modulus increased	Cross-linking by N_2 elimination

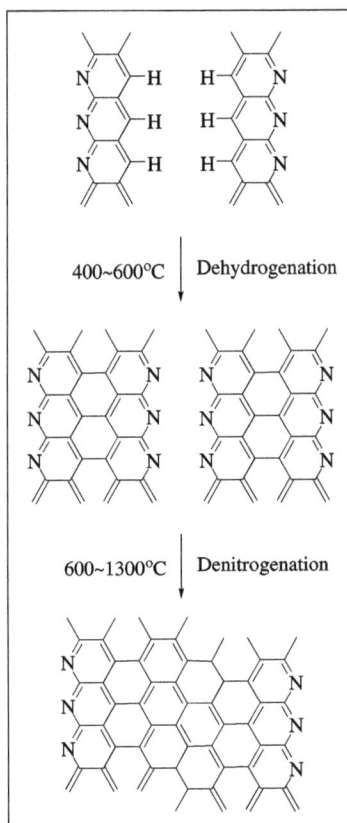

Schematic of graphite structure.

4. Surface treatment and washing: Generally, the surface treatment of carbon fibers was performed to improve the mechanical properties of the composite through alteration of the fiber surface. In many companies, the treatment method of the surface of carbon fibers is still kept confidential. The most often used surface treatment methods for carbon fibers could be categorized as liquid and gaseous oxidation treatments. The liquid oxidation treatment is well known and could double the composite shear strengths with slight reductions (4–6 %) in fiber tensile strengths.

Among the oxidation treatment methods of the liquid type, the anodic oxidation treatment method has been widely used in the surface treatment of commercial carbon fibers because it is inexpensive, fast, and efficient. Figure shows the schematic of the surface treatment and washing baths. In this method, Faraday's Law applies and 96,500 C liberates 1 g equivalent of O_2. The duration of the surface treatment is related to the line speed. In addition, the current density as a standard variable is used to control the treatment level per unit length of carbon fiber during the surface treatment, usually expressed as C/m.

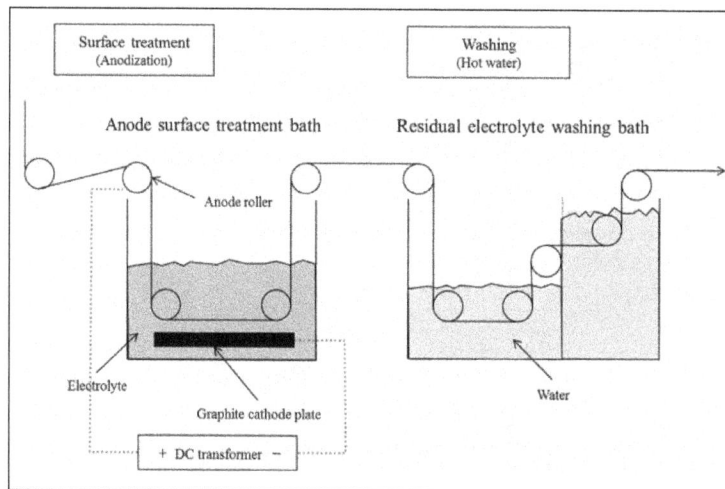

Schematic of surface treatment and washing baths.

The electrically conductive carbon fibers form the anode during the electrolysis of an acid or a salt solution such as nitric acid (HNO_3), sulfuric acid (H_2SO_4), ammonium sulfate (($NH_4)_2SO_4$), and ammonium bicarbonate (NH_4HCO_3). One of the electrolytes, ammonium sulfate, is usually used in the commercial surface treatment processes of carbon fibers. This causes the carbonyl containing groups such as COOH to form on the smooth fiber surface. The carbonyl groups improve the cohesion between the fiber and resin used in the final composite. After this surface treatment, the excess electrolyte is removed using warm water wash treatment. The carbon fibers are then passed onto the next process through one or more water baths constantly flowing with water.

Currently, the demand for the surface treatment of carbon fibers has significantly increased with increasing necessity for high-performance carbon fiber composites.

5. Drying, sizing, and winding: Carbon fibers require some protection or lubrication for the ease of handling because of their brittleness. The carbon fibers are predried for the sizing treatment, and the sizing materials are selected such that they protect the physical characteristics of carbon fibers. These sizing materials need to provide consistent handling and not build up residue on the processing equipment. The sizing materials also need to be compatible with the matrix resin. This includes solubility in and reactivity with the formulated resin. This allows the resin to penetrate

the fiber bundle and interact with the fiber surface. Generally, the epoxy resins or epoxy formulations are used as sizing materials. The sizing materials should not change either the chemical or physical characteristics of the carbon fibers during storage. Some sizing materials are water soluble and washable after either weaving or braiding.

The fiber sizing, process to apply sizing, and sizing content are considered to be critical factors in the carbon fiber specification. The type of size material and particle size of the aqueous dispersion must be controlled to establish good properties in the carbon fibers after sizing. From these viewpoints, the types of emulsifier and resin and their respective concentrations are key to improving the characteristics of the carbon fibers. The control of wetting in the sizing bath is needed to control the level of size on the carbon fibers. All the steps pertaining to the application of the sizing to the carbon fiber and drying must also be consistent.

Many sizing materials such as epoxy resins are not soluble in water and must be applied as a dispersion of emulsion in water. This could result in the sizing being uniformly distributed on the surface of the fibers. Alternately, the sizing materials could exist as either droplet on the fiber surface or by sticking together a number of individual fibers. The particle size of the emulsion in the sizing bath is controlled to provide a dependable product.

Meanwhile, the dried carbon fibers after the sizing treatment are collected using winders. The winding machines generally operate automatically. In addition, the winders could usually produce finished spools of up to 12 kg in weight.

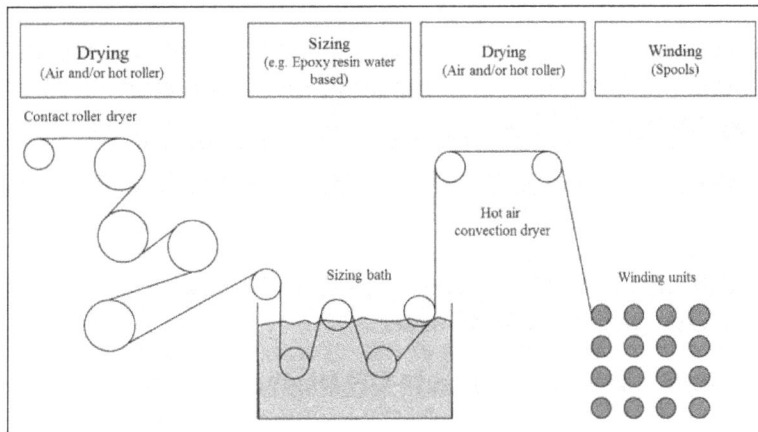

Schematic of drying, sizing, and winding.

Types of Polyacrylonitrile-based Carbon Fibers

There are three basic categories of PAN-based carbon fibers:

- Large tow: Inexpensive to manufacture can be conveniently chopped to a staple form.

- General purpose grades: Less stringent product qualification.

- Aerospace grades: Premium-grade products.

Each category of carbon fibers is available in several production types, which are based on their tensile strength and modulus:

- General purpose

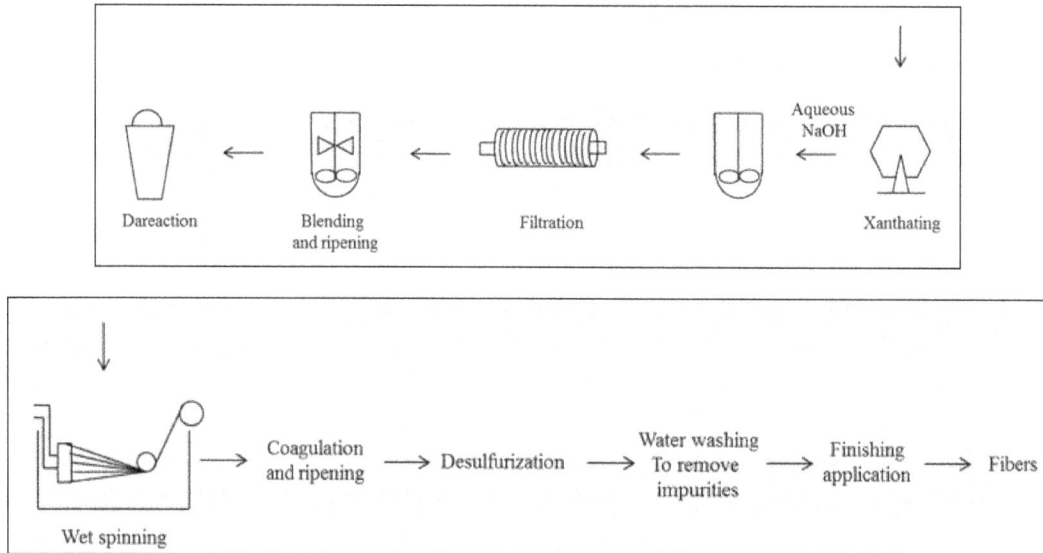

Process of manufacture of viscose rayon fibers.

2. Pressing: The swollen alkali cellulose mass is pressed to a wet weight equivalent of 2.5–3.0 times the original pulp weight to obtain an accurate ratio of alkali to cellulose.

3. Shredding: The pressed alkali cellulose is shredded mechanically to yield finely divided, fluffy particles called "crumbs." This step provides increased surface area of the alkali cellulose, thereby increasing its ability to react in the steps that follow.

4. Aging: The alkali cellulose is aged under controlled conditions of time and temperature (between 18 and 30 °C) in order to depolymerize the cellulose to the desired degree of polymerization. In this step, the average molecular weight of the original pulp is reduced by a factor of two to three. The reduction of the cellulose is done to get a viscose solution of right viscosity and cellulose concentration.

5. Xanthation: In this step, the aged alkali cellulose crumbs are placed in vats and are allowed to react with carbon disulfide under controlled temperature (20–30 °C) to form cellulose xanthate.

$$(C_6H_9O_4ONa)_n + nCS_2 \longrightarrow (C_6H_9O_4 - SC - SNa)_n$$

Side reactions, which occur along with the conversion of alkali cellulose to cellulose xanthate, are responsible for the orange color of the xanthate crumb as well as the resulting viscose solution. The orange cellulose xanthate crumb is dissolved in dilute sodium hydroxide at 15–20 °C under high-shear mixing conditions to obtain a viscous orange colored solution called "viscose," which is the basis for the manufacturing process. The viscose solution is then filtered (to remove the insoluble fiber material) and deaerated.

6. Dissolving: The yellow crumb is dissolved in an aqueous caustic solution. The large xanthate substituents on cellulose force the chains apart, reducing the interchain hydrogen bonds and allowing water molecules to solvate and separate the chains, thereby leading to a solution of the otherwise insoluble cellulose. Because of the blocks of the unxanthated cellulose in the crystalline regions, the yellow crumb is not completely soluble at this stage. Because the cellulose xanthate solution (or more accurately, suspension) has high viscosity, it has been termed "viscose."

7. Ripening: The viscose is allowed to stand for a period of time to "ripen." Two important processes occur during ripening: redistribution and loss of xanthate groups. The reversible xanthation reaction allows some of the xanthate groups to revert to the cellulosic hydroxyls and free CS_2. This free CS_2 can then either escape or react with other hydroxyls on other portions of the cellulose chain. In this way, the ordered, or crystalline, regions are gradually broken down, and a more complete solution is achieved. The CS_2, which is lost, reduces the solubility of the cellulose and facilitates the regeneration of the cellulose after it is formed into a filament.

$$(C_6H_9O_4SC-SNa)_n + nH_2O \rightarrow (C_6H_{10}O_5)_n + nCS_2 + nNaOH$$

8. Filtering: The viscose is filtered to remove the undissolved materials, which might either disrupt the spinning process or cause defects in the rayon filament.

9. Degassing: Bubbles of air entrapped in the viscose must be removed prior to extrusion, or else, they cause voids, or weak spots, in the fine rayon filaments.

10. Spinning (Wet Spinning): Production of Viscose Rayon Filament - The viscose solution is metered through a spinneret into a spin bath containing sulfuric acid (to acidify the sodium cellulose xanthate), sodium sulfate (to impart a high salt content to the bath, which is useful in the rapid coagulation of viscose), and zinc sulfate (for exchange with sodium xanthate to form zinc xanthate, to cross-link the cellulose molecules). Once the cellulose xanthate is neutralized and acidified, a rapid coagulation of the rayon filaments occurs, followed by simultaneous stretching and decomposition of cellulose xanthate to regenerated cellulose. The stretching and decomposition processes are vital for obtaining the desired tenacity and other properties of rayon. The slow regeneration of cellulose and stretching of rayon leads to greater areas of crystallinity within the fiber, as is done with high-tenacity rayon.

The dilute sulfuric acid decomposes the xanthate and regenerates cellulose by wet spinning. The outer portion of the xanthate is decomposed in the acid bath, forming a cellulose skin on the fiber. Sodium and zinc sulfates control the rate of decomposition (of cellulose xanthate to cellulose) and fiber formation.

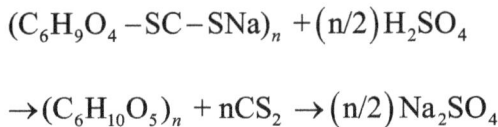

$$(C_6H_9O_4 - SC - SNa)_n + (n/2)H_2SO_4$$

$$\rightarrow (C_6H_{10}O_5)_n + nCS_2 \rightarrow (n/2)Na_2SO_4$$

Elongation-at-break is seen to decrease with increasing degree of crystallinity and orientation of rayon.

11. Drawing: The rayon filaments are stretched, while the cellulose chains are still relatively mobile. This causes the chains to stretch out and orient along the fiber axis. As the chains become more parallel, the interchain hydrogen bonds form, providing the filaments with the properties necessary for use as textile fibers.

12. Washing: The freshly regenerated rayon contains many salts and other water soluble impurities, which need to be removed. Several different washing techniques may be used.

13. Cutting: If the rayon is to be used as staple (i.e., discreet lengths of fiber), the group of filaments (termed "tow") is passed through a rotary cutter to yield a fiber, which can be processed analogously to cotton.

Currammonium Rayon

In this process, the fibers are produced in a solution of cellulosic material in cuprammonium hydroxide at a low temperature in nitrogen atmosphere, followed by extruding through a spinneret into a sulfuric acid solution, necessary to decompose the cuprammonium complex to cellulose. This is a relatively more expensive process than that used for viscose rayon. However, the cross sections of the resultant fibers are almost round.

Saponified Cellulose Acetate Rayon

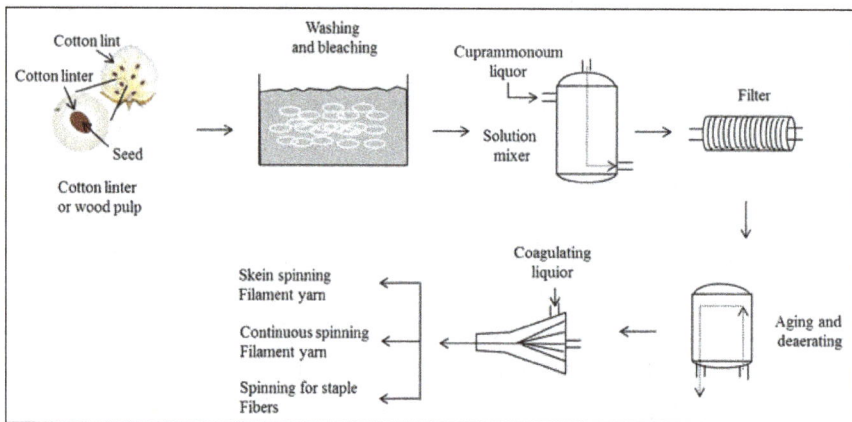

Process of manufacture of currammonium rayon fibers.

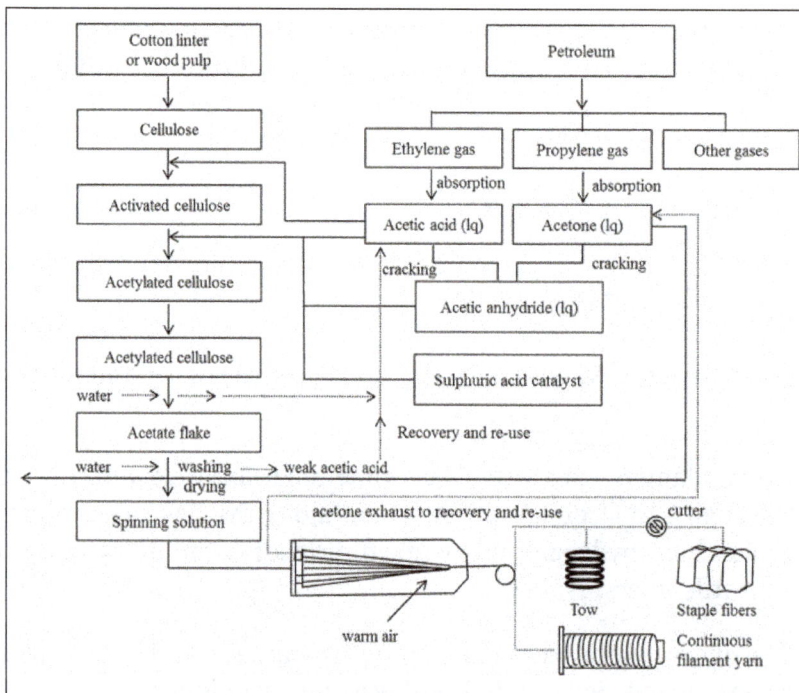

Process of manufacture of saponified cellulose acetate rayon.

Rayon can be produced from cellulose acetate yarns through saponification. Purified cotton is steeped in glacial acetic acid to make it more reactive. It is then acetylated with an excess of glacial acetic acid and acetic anhydride, followed by sulfuric acid to promote the reaction. The cellulose triacetate formed by acetylation is hydrolyzed to convert the triacetate to diacetate. The resultant mixture is poured into water, which precipitates the cellulose acetate. For spinning, it is dissolved in acetone, filtered, deaerated, and extruded into hot air, which evaporates the solvent. A high degree of orientation can be given to the fiber by drawing because cellulose acetate is more plastic. Its fiber cross section is nearly round, but lobed.

Manufacture of Carbon Fibers from Cellulosic Precursors

Rayon fibers, as mentioned above, can be converted into carbon fibers with chemical, physical, mechanical, and microstructural changes, through stabilization and carbonization processes. With regard to chemical and mechanical properties, the stabilization of the precursor fibers is important to produce stable carbon fibers through the subsequent carbonization process. In particular, the thermal shrinkage of the cellulose fibers occurs owing to the weight reduction of fibers during the stabilization process. Therefore, in this section, only the stabilization, carbonization, and graphitization of cellulosic precursors for manufacturing of carbon fibers will be described. Other processes such as surface treatment, sizing, and winding are identical to the processing of the PAN-based carbon fibers.

Stabilization

Cellulose is a glucose-based, linear polymer connected by β-(1-4) glycosidic linkages. The hydroxyl groups can easily form intra- and intermolecular hydrogen bonds in the cellulose structure, and the hydrogen bonds can lead to various ordered crystalline arrangements. From the molecular stoichiometry of $(C_6H_{10}O_5)n$, it was found that the theoretical carbon yield of the carbonization process for the cellulosic structure is 44.4 %. However, the actual carbon yield is only between 10 % and 30 %, resulting from the depolymerization of the macromolecular chains and the elimination of carbon by oxygen in the forms of carbon monoxide (CO), carbon dioxide (CO_2), aldehydes, organic acids, and tars. Therefore, the selection of suitable stabilizer materials is essential for improving the carbon yield and properties of the products in the carbonization process, which can lead to higher efficiency.

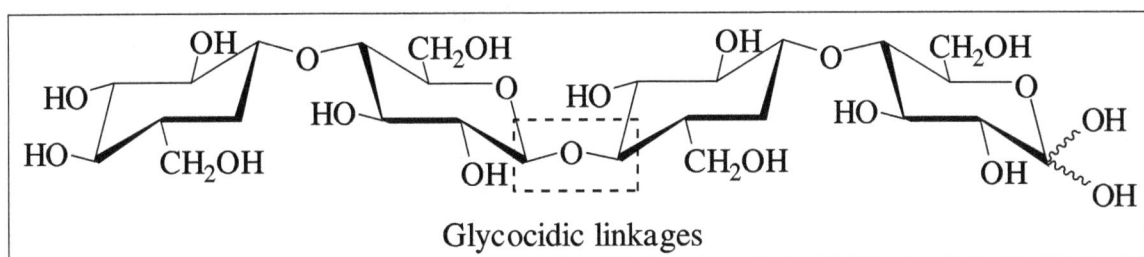

Glycocidic linkages

Chemical structure of cellulose polymer chain.

The degradation of general cellulose fibers under an inert atmosphere initiates at 200 °C and ends at around 380 °C. In addition, the thermal stability of the cellulose sources depends on the process conditions for cellulose fiber production. Although the physicochemical processes taking place during the transformation of cellulose into carbon are complex, it is certain that the depolymerization of the macromolecular chains produces a variety of oxygenated compounds. This leads to the major mass loss of the solid residue through the production of volatile substances. Various methods can help in reducing the burning loss. One method involves the modification of cellulose precursors to improve the yield and properties of the carbon fibers. The other method involves the pyrolysis of the cellulose with slow heating rates of a few °C/h or the treatment of cellulose fibers with suitable impregnators.

The pyrolysis of cellulose is mainly controlled by two predominant reactions, dehydration and depolymerization (cleavage). The main reaction at low temperatures (300 °C) is the dehydration reaction for the stabilization of the cellulose structure. During the dehydration, the elimination of the hydroxyl groups leads to conjugated double bonds, and subsequently, the dehydrated cellulose ring becomes less accessible for cleavage compared to the original cellulose structure in an aromatic form in the carbonization step. The polymeric structure is basically retained through dehydration, and the weight loss at this temperature range is usually owing to the evaporation of water. Depolymerization in the early stages of pyrolysis with the incomplete dehydration of the cellulose structure causes major mass loss at higher temperatures. Therefore, a slow heating rate of a few °C/h can be used to increase the carbon yield of the cellulose fibers. In addition, the slow pyrolysis influences the properties of the final fibers such as improved density, porosity, and microstructure, compared to fast pyrolysis.

Carbonization and Graphitization

The first H_2O dehydration occurs between 25 and 150 °C (Stage I), and the physical desorption of water and dehydration of the cellulosic unit occur between 150 and 240 °C (Stage II). This leads to the formation of the double-bonded intermediates. Carbonization of cellulose refers to the conversion process from this depolymerized structure into graphite-like layers through repolymerization. The process begins at approximately 300 °C and continues up to 900 °C, as shown in the Stages III and IV of figure. The basic microstructure of the carbon is formed during the Stage III. As shown in figure. the thermal cleavage of the glycosidic linkage and the scission of ether bonds occur over the range 240–400 °C. Moreover, depolymerization to monosaccharide derivatives occurs during this stage of carbonization. Then, these intermediates form aromatic structures, releasing gases containing non-carbon atoms (O, H). The carbonaceous residue is converted into a more ordered carbon structure through heat treatment between 400 and 900 °C under an inert atmosphere. The full mechanism of aromatization related to the graphitic products is still unknown, owing to the complex characterization of the cellulose decomposition and existence of many competing reactions during the different stages of carbonization.

The heat treatment up to 900 °C causes the formation of semiordered carbonaceous structures under an inert atmosphere. After this stage, the carbonized fibers can be treated with heat at higher temperatures to initiate graphitization. In general, graphitization is carried out under stress at 900–3,000 °C to obtain high-modulus fibers through the development of an enhanced order of the graphene stacks, both laterally between the layers (crystallographic register) and in terms of the preferred orientation along the fiber axis. The Young's modulus is increased by the increasing treatment temperature, if the graphitization is conducted under tension. After graphitization, the carbon content of the fibers usually increases to above 99 %, and the fiber density increases resulting from the growth of crystallites.

Pitch Precursors

Pitches are complex blends of polyaromatic molecules and heterocyclic compounds, which can be used as precursors of carbon fibers or carbon fillers in carbon composites. These pitches can contain more than 80 % carbon, and the composition of a pitch varies with the source tar and processing conditions. In addition, these pitches can be obtained from one of several sources mentioned below:

1. Petroleum refining, normally called bitumen, or asphalt.

2. Destructive distillation of coal.

3. Natural asphalt, e.g., from Trinidad.

4. Pyrolysis of PVC.

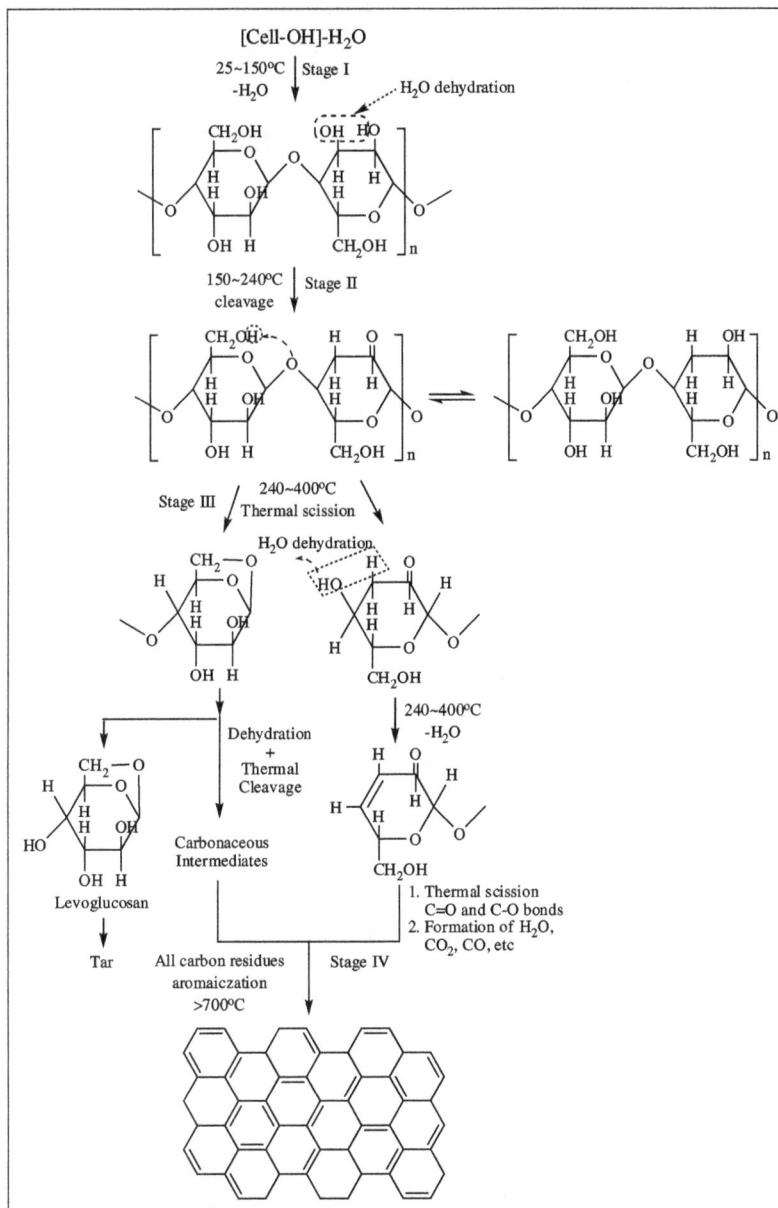

Reactions involved in conversion of cellulose into carbon fibers.

Table: Compositions of various oils and pitches.

Compound	Asphaltene (%)	Polar aromatic (%)	Naphthene aromatic (%)	Saturate (%)	Softening point (°C)
Carbon black oil	2.5	10.6	69.0	17.9	
EXXON (DAU) bottoms (refinery sludge)	14.5	41.1	18.1	26.3	29
Ashland 240 petroleum pitch	64.4	8.6	25.4	1.6	119
Ashland 260 petroleum pitch	82.7	5.9	11.4	0.0	177

The natural pitches are produced by the refining of petroleum and destructive distillation of coal, while the synthetic pitches are produced by the pyrolysis of synthetic polymers. Generally, among the prepared pitches using several sources, the petroleum pitch and coal pitch are widely used in the production of carbon fibers.

In terms of the components of pitch, Riggs et al. considered that the pitch is composed of four main classes of chemical compounds:

1. Saturates: Low molecular weight aliphatic compounds.

2. Naphthene aromatics: Low molecular weight aromatics and saturated ring structures.

3. Polar aromatics: Higher molecular weight and more heterocyclic in nature.

4. Asphaltenes: Highest molecular weight fraction in pitch with the highest aromaticity and thermally most stable.

The compositions of various oils and pitches are listed in table Several researchers have confirmed that the asphaltene rich materials are the most suitable for conversion into carbon fibers.

Petroleum Pitch Precursors

Petroleum pitch can be obtained from various sources such as heavy residue obtained from a catalytic cracking process and steam cracker tar—a by-product of the steam cracking of naphtha or gas oils to produce ethylene or any residues from crude oil distillation or refining. Many methods can be used for the production of pitch and are based on an initial refining process, which can include either one method or combination of several treatment methods listed below:

1. Prolonged heat treatment to advance the molecular weight of the components.

2. Air bowing at approximately 250 °C.

3. Steam stripping and application of vacuum to remove low boiling components.

4. Distillation.

In common with the coal tar pitch, the chemical and physical characteristics of petroleum pitch are dependent on the process and conditions employed especially the process temperature and heat treatment time. Generally, longer times and higher temperatures produce pitches with increased aromaticity and higher anisotropic contents. The petroleum pitches are usually less aromatic compared to the coal tar pitch.

Coal Tar Pitch Precursors

Coal tar is a by-product of the coking of bituminous coals to produce cokes. The metallurgical cokes are produced at high temperatures (between 900 and 1,100 °C), but it produces smokeless fuels at lower temperatures (approximately 600 °C). The low-temperature process affords a smaller amount of tar compared to the high-temperature process. Coal tar pitch is obtained from the coal tar using distillation and heat treatment processes. The pitch is the residue, which follows the removal of the heavy (creosote or anthracene) oil fractions. The pitches are complex mixtures containing many different individual organic compounds, and the precise compositions and properties vary according to the source of the tar and method of removal of low molecular weight fractions.

Roughly two-thirds of the compounds isolated hitherto from the coal tar pitch are aromatic, and the rest are heterocyclic. Most of the compounds are substituted with the methyl group. The majority of the coal tar pitch components contain between three and six rings, having boiling points over the range 340–550 °C. Table lists the major aromatic hydrocarbon compounds, which have been quantitatively found in a typical coal tar pitch.

Preparation Methods of Pitch-based Precursors

Preparation Methods of Petroleum Pitch

Many petroleum products are referred to as "pitch" by the petroleum industry. This could cause considerable confusion outside the refining community. In most cases, the different types of petroleum pitch exist as black solids at room temperature. The individual characteristics of the petroleum pitches vary with the functions of the feedstock and specific processes used in their manufacture. Feedstock can range from being predominantly aliphatic to predominantly aromatic-type chemical structures. A reaction step is used to generate and concentrate the large molecules typically observed in petroleum pitch. The most common processes used to generate petroleum pitches are either singular processes or a combination of: (a) solvent deasphalting, (b) oxidation, and (c) thermal processes.

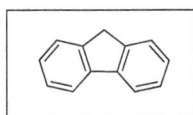	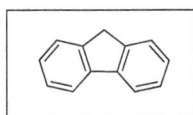	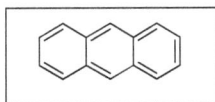	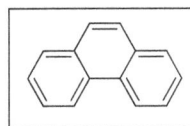	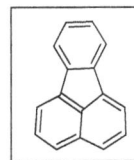
Acenaphthene	Fluorene	Anthracene	Phenanthren	Fluoranthene

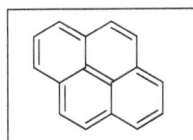	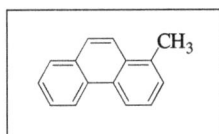	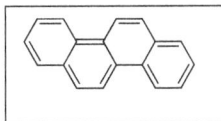	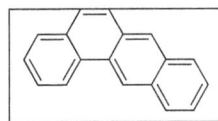	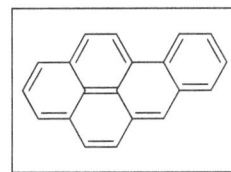
Pyrene	Methylphenanthrene	Chrysene	Benz-(a)-anthracene	Benzo-(a)-pyrene

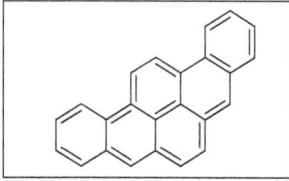

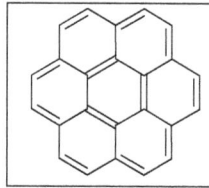

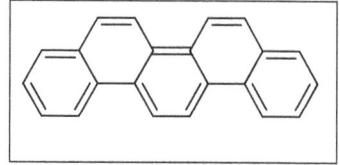

| 3, 4, 8,10-Dibenzopyrene | Coronene | Picene |

Solvent deasphalting is used to separate the fractions of various heavy oils. It involves mixing the feedstock with a paraffinic solvent such as propane, butane, or pentane. The mixing of the feedstock with these light paraffinic solvents causes the precipitation of molecules with higher molecular weights and aromaticities. The chemical and physical properties of this type of petroleum pitch are more closely associated with the asphalt cements used for paving roads. Typical properties include specific gravity of approximately 1.0 g/cc at 60 °F, with the chemical composition containing significant amounts of nonaromatic hydrocarbons and high levels of iron, nickel, and vanadium.

Various grades of pitch can be produced by the oxidation of heavy petroleum hydrocarbons. Although oxygen is used in the process, the products typically do not contain significant amounts of oxygen. During this reaction, the presence of oxygen is successful in generating free radicals, which induce the polymerization reactions. The chemical properties of these products will depend upon the starting material and degree of reaction, but the pitches produced typically have low coking values and high viscosities.

Thermal processing is used to produce the petroleum pitch, as noted in several patents. The thermal processes typically employ heat treatment temperatures over the range 300–480 °C. A typical schematic for producing the petroleum pitch from crude oil using thermal processing is shown in figure.

Preparation Methods of Coal Tar Pitch

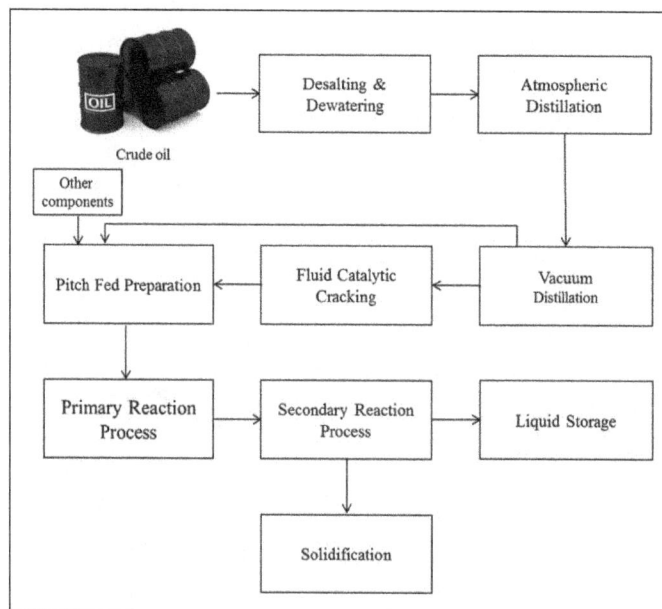

Schematic for the manufacturing process of petroleum pitch from crude oil.

Coal tar is a by-product of the coking of coal, used to produce metallurgical coke. Coal is heated to approximately 1,100 °C in a coke oven to produce coke (the primary product) and by-products such as coke oven gas, coal tar light oil, and coal tar. The typical yields are 70 % solid products and 30 % liquid products. The yield of coal tar, the feedstock for producing coal tar pitch, from a ton of coal is 30–45 L (8–12 gallons). Coal tar pitch has several uses, but the majority of the pitch produced is used as a binder for petroleum coke to produce anodes and graphite electrodes. Figure shows the schematic for coal tar pitch production. As indicated in the figure, the coal tar pitch is produced by the distillation of coal tar.

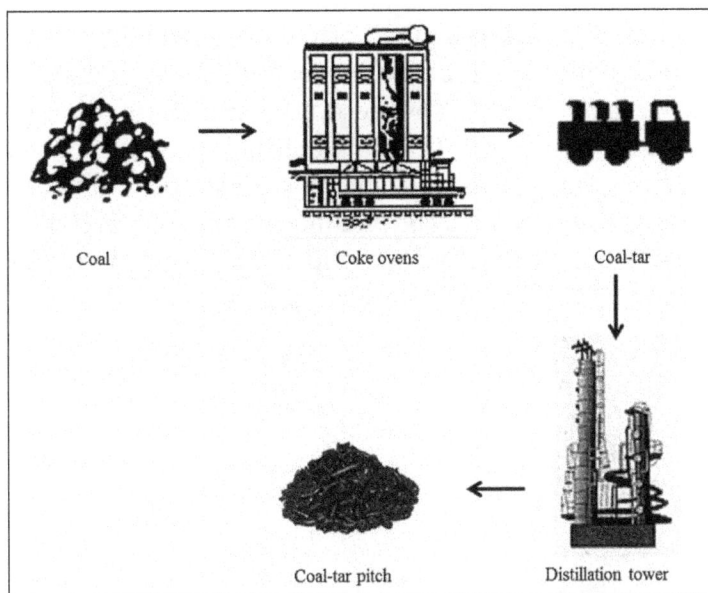

Schematic for the manufacturing process of coal tar pitch.

Pitch as a precursor has the advantage of lower material cost, higher char yield, and higher degree of orientation compared to those of PAN. The graphitic structure also affords pitch-based carbon fibers, higher elastic modulus, and higher thermal and electrical conductivity along the fiber direction. However, the processing cost (mainly pitch purification, mesophase formation, and fiber spinning) to achieve high-performance carbon fibers is higher. The pitch from petroleum and coal tar is isotropic. By evaporating the low molecular weight fractions, the isotropic pitch can be melt-spun into low-cost general purpose (GP) (low strength and low modulus) carbon fibers. To produce high-performance fibers, an expensive hot stretching process needs to be applied. A more common method to produce high-performance carbon fibers from the pitch is to use an anisotropic pitch such as mesophase pitch.

Isotropic Pitch

Isotropic pitches are used to make a GP grade of pitch carbon fibers, which are not graphitic and have poorer properties compared to the high-performance grade (HP), which requires a special treatment process to convert the pitch to a mesophase grade, i.e., an optically anisotropic and graphitic material. The isotropic pitch has to be treated to generate a product suitable for melt spinning, with low volatility and filtered to remove the solid particles. A good starting material would be Ashland 240, which has low quinoline content. The volatile components can be effectively removed with a wiped film evaporator, wherein a thin film of the molten pitch in the evaporator is continuously wiped over the heating surface, exposing a fresh surface, thus permitting the efficient removal of some of the volatile components without overheating. Also, while the pitch is molten, any solid impurities are removed by filtration. These refining processes raise the softening point and avoid the formation of mesophase. Hence, increasingly, pitches such as Ashland's Aerocarb 60 and 70 can be used. The preparations of isotropic pitches have been undertaken to make GP carbon fibers.

Mesophase Pitch

The mesophase pitches used for high-modulus carbon fiber production can be formed either by the thermal polymerization of petroleum-or coal-tar-based pitches, or the catalytic polymerization of pure compounds such as naphthalene. The mesophase transformation results in an intermediate phase, formed between 400 and 550 °C, by the thermal treatment of aromatic hydrocarbons. During the mesophase formation, domains of highly parallel, plate-like molecules form and coalesce until 100 % anisotropic material may be obtained in due course. It has been well established that when the mesophase pitch is carbonized, the morphology of the pitch is the primary factor that determines the microstructure of the resulting graphitic material.

The typical methods used for the production of the mesophase pitch are listed below:

(1) Pyrolysis of isotropic pitch

(2) Solvent extraction

(3) Hydrogenation

(4) Catalytic modification

Manufacture of Carbon Fibers from Pitch-based Precursors

The production of carbon fibers from pitch-based precursors proceeds in several stages: production of precursor fibers, stabilization of the precursor fibers, carbonization of the stabilized precursor fibers, and graphitization of the carbonized precursor fibers. The stabilization process is the most important stage during which the thermal oxidation takes place in the precursor carbon fibers. Only properly stabilized precursor fibers can assure satisfactory performance of the final carbon fiber product. Other processes such as surface treatment, sizing, and winding are identical to the processing of PAN-based carbon fibers.

The typical manufacturing steps for the production of carbon fibers from pitch-based precursors are listed below:

1. Production of precursor fibers by melt spinning: Melt spinning involves three steps: melting the precursor, extrusion through a spinneret capillary, and drawing the fibers as they cool. Many investigators have concluded that this process is the primary source of structure in the mesophase pitch-based carbon fibers and that thermal treatment only reinforces this structure. The effects of melt spinning conditions on the structure and properties of carbon fibers will be discussed in this topic.

Several process variables are important in determining the fibers' potential for developing an ordered graphitic structure. The first is melt temperature. Each mesophase has a range of temperature over which the melt spinning is possible. Spinning at temperatures below this range results in high viscosities and brittle fracture during drawdown, while at temperatures above this range, thermal degradation of the pitch and dripping owing to low viscosity occur. The viscosities of all the mesophases are highly dependent upon temperature, as shown in figure therefore, the temperature range for successful melt spinning is quite narrow. Using the AR mesophase, Mochida found that decreasing the melt spinning temperature by even 15° resulted in a fourfold increase in viscosity and, in turn, significant decreases in the tensile strength and Young's modulus. It was, therefore, concluded that a low melt viscosity was necessary for the production of high-quality carbon fibers.

2. Stabilization of precursor fibers: Analogous to the PAN precursor fibers, the pitch fibers are either infusibilized or oxidized in air at elevated temperatures before being exposed to the final high temperature carbonization treatment. The oxidization temperature should be below the fiber softening point to keep the orientated structure. Depending on the composition, the mesophase pitch precursor is stabilized in air at 250–350 °C for a duration ranging from 30 min to several hours. There has been no consensus hitherto on the purpose of the fiber stretching in this step. The oxidized pitch molecules contain ketone, carbonyl, and carboxyl groups, which lead to the stronger hydrogen bonding between the adjacent molecules.

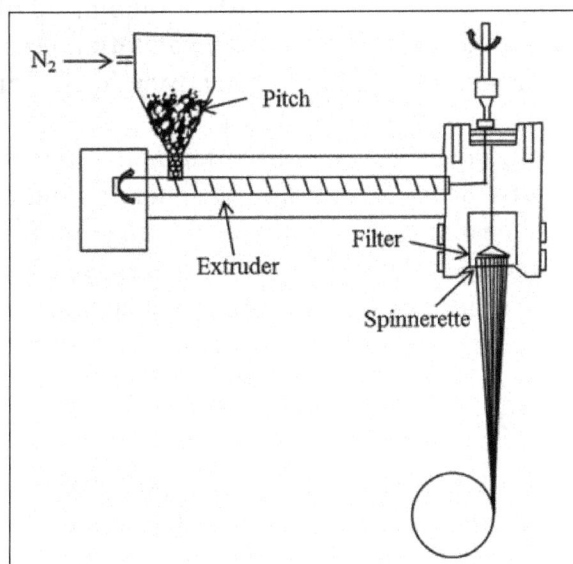

Schematic of melt spinning process for production of precursor fibers.

The introduction of oxygen containing groups and formation of hydrogen bonding between the molecules facilitate the three dimensional cross-linking, but hinder the growth of crystallites. Iodine has been used to reduce the stabilization time and increase the carbon yield of carbon fibers from the natural pitch. In a patent by Sasaki and Sawaki, the pitch fiber was soaked in a methanol solution of iodine till at least 0.05 wt. % of iodine was imbibed. The fiber was then heated under an oxidizing atmosphere for infusibilization. The infusibilization time was affected by the amount of imbibed iodine but generally could be completed within approximately 10 min.

3. Carbonization and graphitization of stabilized precursor fibers: Stabilized fibers are then carbonized and graphitized. The greatest weight loss takes place in the early stages of carbonization.

In order to avoid the defects created by the excessive release of volatiles, the fibers are precarbonized for a brief period of 0.5–5 min at 700–900 °C. The carbon fibers can be produced by carbonizing the stabilized fibers to 1,500–1,800 °C. Bright and Singer reported that because of the degradation of the structure, the modulus decreased at temperatures up to approximately 1,000 °C, but increased significantly upon further increase in temperature. The carbon fibers can be graphitized at temperatures close to 3,000 °C for enhanced Young's modulus. Barr et al. have shown that increasing the heat treatment temperature could increase the preferred alignment of the crystalline lamellae.

Although graphite layers are aligned along the fiber axis, the transverse structures of the carbon fibers can be different. The velocity gradients orient the layers radially, circumferentially, or randomly. It has been reported that a radial crack can form in the mesophase carbon fibers with layer planes distributed radially. The alignments in the precursor fiber are retained in the resultant carbon fiber. Therefore, carbon fiber strength could be improved by adjusting the microstructure in the precursor fiber. Research has shown that the flaw sensitivity of mesophase carbon fibers is reduced by varying the microstructure of the pitch precursor fibers. The microstructure can be modified by changing the flow profiles during melt spinning. A radial cross section is usually formed through the laminar flow of the pitch melt. Petoca Oil Company has employed agitation in the spinneret to impart a randomized distribution for the folded graphene layer planes in the transverse direction. The agitation created a turbulent flow and the resultant carbon fibers showed increased tensile strength. The turbulent flow can also be obtained by different die designs because the flow behavior heavily depends on the shape of the spinneret. Spinnerets containing sections with different diameters have shown to be able to change the fiber microstructure. Because melt flow is dependent on melt viscosity, the change in the microstructure can also be obtained simply by changing the spinning temperature. Otani and Oya have shown that when the spinning temperature was raised to above 349 °C, the radial structure changed to either a random type or radial type, surrounded by an onion-skin-type structure.

Other Forms of Precursors

Many other polymers have also been investigated for their potential as carbon fiber precursors. In addition to the cellulosic fibers, other natural fibers have been investigated and considered as the precursors for carbon fibers such as silk, chitosan, and eucalyptus. They can lower the production cost; however, most of them are used for GP carbon fibers, which do not afford strong mechanical properties.

In addition, many linear and cyclic polymers have been investigated as a precursor for the production of carbon fibers, including phenolic polymers, polyacenephthalene, polyamide, polyphenylene, poly-p-phenylene benzobisthiazole (PBBT), polybenzoxazole, polybenzimidazole, polyvinyl alcohol, polyvinylidene chloride, and polystyrene. Linear precursors require heat stretching to obtain high-performance carbon fibers, and their carbon yields are usually extremely low. The polymers with high aromatic content can generally offer a high carbon yield and in some cases easy stabilization. However, these polymers either have high costs or do not produce high-performance carbon fibers. Further research needs to be conducted to reduce the processing cost while improving the mechanical properties of the resultant carbon fibers.

Processing of Carbon-Carbon Composites

Fabrication Process

Classically, carbon ceramics are fabricated by combining solid particles of pure carbon (known as primary carbon) with a preliminary binder which acts as a precursor for secondary carbon formed during the carbonisation process. In CC composites the carbon fibres (based on rayon/PAN/pitch) in the form of U-D, 2-D and multi-directional preforms are used as primary carbon rather than particulate fillers. There are two distinct techniques used to fill the interstices between the carbon fibres. These are (i) gas phase using a chemical vapour deposition process, and (ii) the liquid phase route using thermosetting resins or pitch (PIC). The fibres can be very stiff, highly oriented graphitic and dense (pitch precursor) or relatively flexible, stronger, less oriented graphitic and less dense (P AN precursor). Conversely, the matrix can be highly oriented and graphitic if produced from pitch, either isotropic or anisotropic if produced from CVO or usually isotropic if produced from thermosetting resin using phenolic resin. A general rule of thumb employed by manufacturers is that the gas phase route is adequate for thin-walled parts and the liquid phase route is preferable for thick parts. A combination of liquid and gas phase processes is also being followed.

Multidirectionally Reinforced Preforms

The main advantage of multidirectional CC composites is the freedom to orient selected fibres and amounts to accommodate the design loads of the final structural component and make them virtually delaminatian free. Multidirectional preform fabrication technology provides the means to produce tailored and near net shape composites, which meet the directional property requirement of an end item.

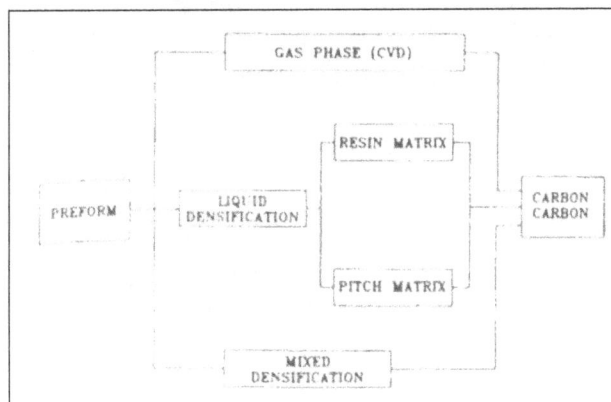

Densification by carbon matrix.

Thermal, mechanical and physical properties of the composites can be controlled by the appropriate design of substrate parameters such as fibre orientation, volume fraction of fibres in the required direction. Preform weaving technology provides the ideal approach to tailor the structural composites. The simplest type of multidirectional structure is based on a three directional (3-D) orthogonal construction as shown in figure, consisting of multiple yarn bundles located within the structure described in cartesian coordinates. In any direction, fibre bundles are straight in order to obtain the maximum structural capability of fibre. The type of fibre, the number of fibre bullets per site, the fibre bundle spacings, volume fraction distributions, the woven bulk densities characterise the preform.

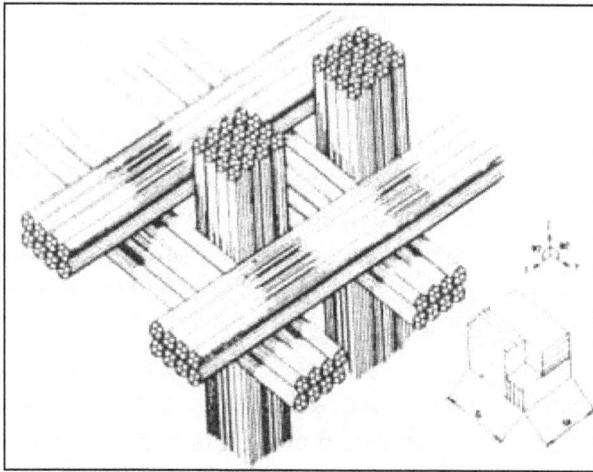

Dimensional Array.

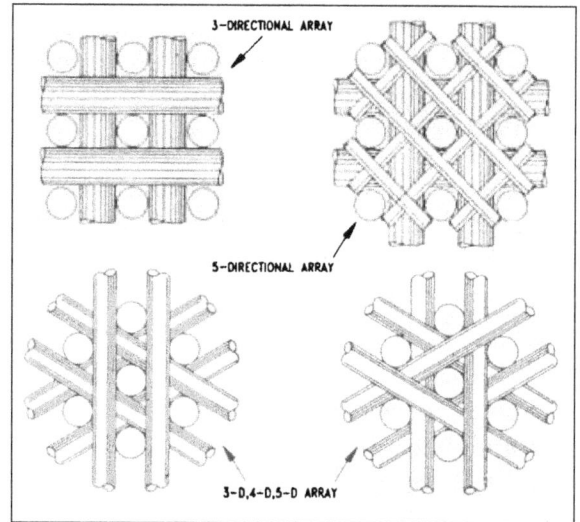

3-D, 4-D and 5-D arrays.

These characteristics are calculated 'for a typical unit cell in the preform. Several weave modifications to the basic 3-0 orthogonal designs are possible as shown in figure, to form a more isotropic structure in 4-D5-D, 7-D and 11-D. To enhance the composite properties between the planes, diagonal yarns are introduced. The multidirectional preform technology, also known as the fibre architecture, employs multidisciplinary approaches of structural engineering, mechanical engineering and textile technology to develop preforms in simple blocks, cylinders, cones, contours, surfaces of revolution and complex geometries and shapes. The techniques employed are conventional weaving with dry yarns, pierced fabrics, assembly of pre-cured rods, on manual, semi-automated and automatic loom set-ups and 3-D braiding and 3-D knitting. Countries like USA, France where this technology was pioneered, have kept this technology closely guarded due to its immediate adaptability to strategic products. With relentless efforts and innovation, OROL has developed the multidirectionally reinforced preform technology for 3-D, 4- D, 5-D and 6-D preforms in blocks and cylinders with varying weave parameters. The technology and facilities are established to develop multidirectional preforms using, manual and semi-automated looms. Figure shows the possible material variants to preforming. The weaving technology and defect figure -dimensional array characterisation techniques are developed to realise defect-free preforms. Efforts are being made to produce preforms in near-net shapes using automation techniques and 3-D braiding technology. Figure shows six different multidirectional preforms woven at DRDL.

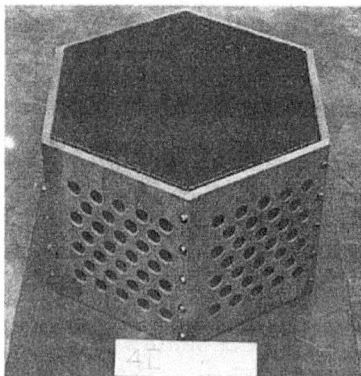

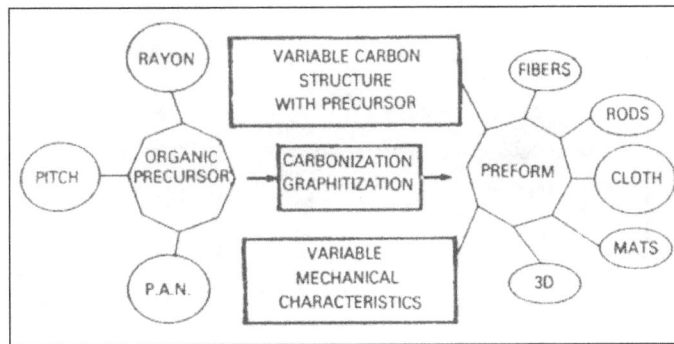

Material variants to preforming.

CC Processing Technology

The CC densification process involves in-depth deposition of secondary carbon from different precursors using either gas phase impregnation or the liquid phase impregnation.

Gas Phase Impregnation (Chemical Vapour Deposition, CVD)

This technique uses volatile hydrocarbons such as methane, propane, benzene and other low molecular weight units as precursors. Thermal decomposition is achieved on the heated surface of the carbon fibre substrates resulting in a pyrolitic carbon deposit. This technique can be employed to deposit carbon on to dry fibre preforms or to densify porous CC structures produced by the liquid impregnation route, in which case it is referred to as chemical vapour infiltration. This process route was widely used by the Western countries for the production of thinner parts like aircraft brake discs and nozzles. CC process technology using CVD technique is yet to be established in our country.

Liquid Phase Impregnation Process

This process involves impregnation with liquid impregnants like coal tar/petroleum pitches and high char-yielding thermosetting resins.

The criterion for selection of impregnates is based on the characteristics like viscosity, carbon yield, matrix microstructure and matrix crystalline structure which are considerably influenced by the time-temperature pressure relationships during the process. The two general categories are aromatic, ring-structured conventional thermosetting resins such as phenolics, furans and advanced resins like ethynyl pyrenes or pitches based on coal tar, petroleum and their blends. Figure 6 shows the CC manufacturing process using the multiple impregnation, carbonisation (1000 °C), high pressure (1000 bars) carbonisation (HIP) and graphitisation (2750 °C).

In atmospheric pressure carbonisation, the carbon yields obtained from pitch are only around 50 per cent i.e. approximating those from high yield thermosetting resins. Yields as high as 90 per cent can be obtained by carbonising the pitch under high pressure of 1000 bars, thus making the process more efficient. Pressure applied during pyrolysis also affects the matrix microstructure. The higher the pressure the more coarse and isotropic will be the microstructure due to the suppression of gas formation and escape. High pressure also helps in lowering the temperature of meso phase formation in pitch, resulting in highly oriented crystalline structure. The HIP process is the only practical route to lower the production cost of CC composites.

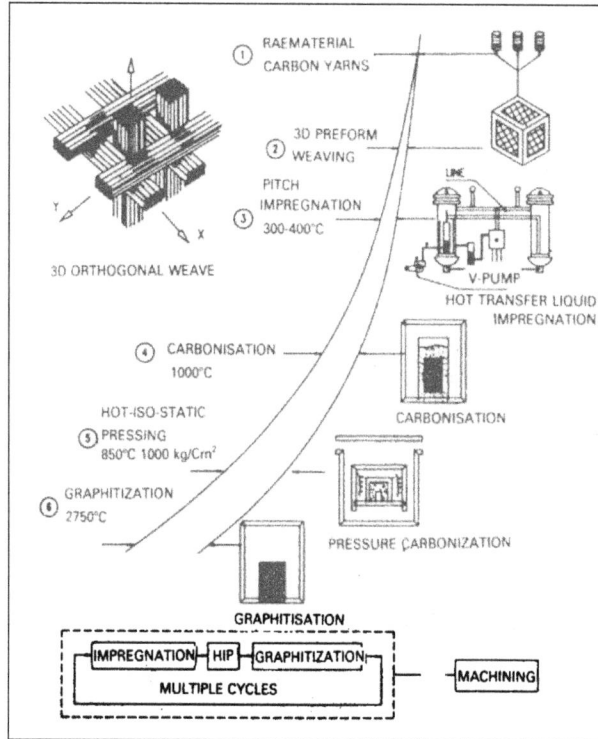

Carbon-carbon manufacturing process.

DRDL has established the state of the art facilities for the prototype production of CC products up to a maximum density of 2.0 g/cc. Apart from the basic process equipments DRDL has designed and fabricated several auxiliary support systems such as centralised nitrogen gas supply, closed loop process cooling station, and ventilation/pollution control devices. Also the CC technology group has established the machining facilities and standardised the machining parameters for CC composites.

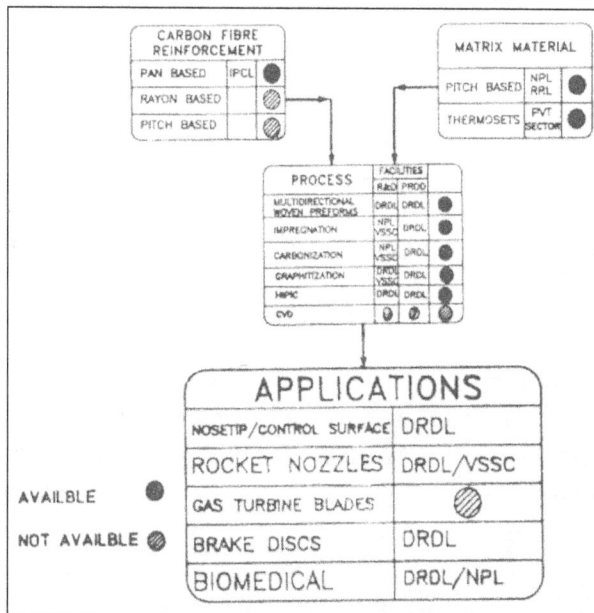

CC composites status in India.

CC process facilities like impregnator/carboniser exist with various institutions like VSSC and NPL. The present status of this technology is depicted in figure.

Oxidation Resistance

Oxidation Protection Mechanisms

Notwithstanding the attractive mechanical and thermal properties of CC at elevated temperatures, some of the potential applications like turbine structural components which require long term exposure 10 high temperature are restricted by the inherent reactivity of carbon towards oxygen beyond 500 °C. A number of different oxidation protection mechanisms have been explored to improve the oxidation resistance of CC composites. The techniques developed can be categorised as:

a) Surface coatings: Single layer/multilayers, using chemical vapour deposition, pack cementation, physical vapour deposition (PVD) and plasma spray.

b) In depth protection includes solgel process, impregnation with inorganic salts (for limited temperature range) and melt impregnation or in depth deposition of SiC matrix.

The materials and process known to give oxidation protection are given in table. With the external protection methods, the thermal expansion mismatch between carbon material and possible refractory coatings is the main problem to be overcome. Microcracks developed in refractory layers have to be sealed with glassy coatings. The best oxidation resistance was achieved in which CVD surface coatings were formed in addition to in depth protection.

Internal protection methods include (i) direct removal and or deactivation of catalytic impurities, and (ii) incorporation of oxidation inhibitors and total or partial substitution of matrix material.

A successful protection system comprises a coating, internal inhibitor and a compatible substrate since CC composites constitute a diverse class of materials with a wide range of mechanical, thermal and morphological properties.

Table: Various materials known to give oxidation protection.

Short-term protection	Long-term protect upto 600 °C
Pack cementation with SiC. SiO_2 followed by impregnation I with alkali silicates to seal the cracks.	Impregnation with inorganic salts, Boron oxides, phosphates, and halogen compounds.
Sintering with SiC and B_4C.	Up 10 1500 °C
Impregnation with tetra ethyl-orthosilicate Solgel process for addition of ceramic powders and glasses.	(i) Chemical deposition of Rh. Tantalum, Carbjdes titanium Silicon.
Chemical vapour	Nitride, tungsten carbide
Spraying of Ni and Si in nitrocellulose lacquer followed by sintering in vacuum to form Ni-Si metallic phase of SiO_2.	(ii) Cermets of refractory materials such as ZrB_2, $MOSi_2$ or Si_3N_4 am carbides, oxides silicides, nitrides of metals like tantalum.
Chemical reaction with molecular silicon to form SiC.	
Hafnium diboride, hafnium' oxide, iridium for temperatures beyond 1700 °C.	

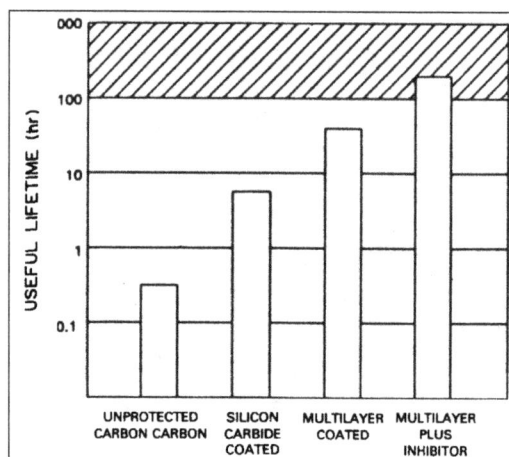

Process in oxidation protection research.

Selection of appropriate fibre, preform fabrication technique, matrix precursor and densification processing method is essential if good oxidation resistance as well as physicochemical compatibility between substrate and coating is to be achieved. The progress on research on oxidation protection is illustrated by the bar chart shown in figure.

Higher Oxidation Resistance

Introduction of a ceramic matrix like SiC instead of carbon matrix in the carbon fibre preform gives higher oxidation resistance than that of oxidation-resistant CC. These composites known as C/SiC composites provide a good tradeoff between the high temperature capability of carbon fibres and the high oxidation resistance of ceramic matrices. Extensive work has been carried out by SEP FRANCE on C/SiC composites for liquid propellant rocket and air breathing engines, thrust vectoring nozzles, hot gas valves and tubes and space plane thermal structures. The third family of thermo-structural composites, viz. SiC/SiC, employ ceramic fibres (SiC) and ceramic matrix (SiC). These composites provide an excellent oxidation resistance for long durations and capable of withstanding thermal cycling for re-usable structures. SiC/SiC composites are used for liquid propellant rocket engine chambers, jet engines, gas turbine components and space thermal structures. However, SiC/SiC composites start losing the mechanical strength beyond 1200 °C unlike carbon carbon composites.

Oxidation Protection of Carbon-Carbon Composites

Developments in science and engineering, beyond imagination, require the implementation and use of high performance material applications with moderate temperature, the polymeric materials have gradually replaced metals. Along with this, Thermostructural composites can replace refractory metal alloys for high temperatures. These materials are able to retain their original properties and the ultra-high temperature. The widely used thermostructural materials are ceramic matrix composites (CMC), metal matrix composites (MMC) and carbon/carbon (C/C) composites. The matrix and the reinforcement in C/C composites is the carbon whose form is very close to graphite and a widely used family thermostructural materials. The interest in these thermostructural materials lies in the interactive properties of each of the three component (fiber, matrix and the interface/interphase). The fibrous reinforcement support all the mechanical load applied to the material. The type of fiber reinforcement and its arrangement is dictated by the final application of the composite. The matrix is the second component and acts as a load balancer, and protection of the reinforcement. The fibrous reinforcement and the matrix are connected by the interphase whose role is to make a non-brittle material out of fragile components.

Carbon/carbon (C/C) composites are widely used in various fields such as aerospace materials, nuclear or Formula I due to their light weight and excellent mechanical and friction properties. The C/C composites with such an excellent mechanical properties, however, have the major drawback to oxidize significantly from relatively low temperatures. This oxidation depends on many parameters such as temperature, partial pressures of oxygen and water vapor and impurities that can act as oxidation catalyst.

Above 370 °C, C/C composites readily oxidize in air, and if CO_2 can escape freely, this will promote the oxidation process.

$$C + O_2 = CO_2$$

The rate of oxidation increases with temperature but decreases with time. An increased composite heat treatment temperature (HTT) will reduce the oxidation rate, probably due to a reduced level of impurities, less edge sites and an overall reduction in carbonization stress. Initially, the fiber matrix interface is attacked, resulting in delamination cracking between plies, but as the temperature increases, more severe oxidation attack occurs within the fiber bundles, promoting cross bundle cracking.

Usually accompanied by a significant deterioration in the mechanical properties, oxidation causes a network of open pores, debonded fibers and the denuded matrix.

This major drawback can be catastrophic if the material is subjected to extreme conditions: high temperatures and large airflows. These can be encountered in some applications of space and aeronautics. This high sensitivity in this type of environment is even more damaging as the consumption rates of the composite by oxygen are extremely high. The development of compounds that can halt or slow this harmful process has long been the subject of numerous studies to protect the C/C composite.

It is, therefore, necessary for many applications at elevated temperatures to apply a protective coating that will provide both oxidation and erosion protection. Basically, there are two modes of protection:

1. Use of inhibitors and sealants to block active sites internally and stifle the rate of oxidation.

2. Apply a barrier coating to form a protective outside layer, preventing O2 moving inwards and carbon moving out.

Surface deposition appears as the most effective in applications at very high temperatures. As part of this report, the focus is on protection against oxidation (PAO) by using different refractory materials. To be very effective, this PAO must be more dense and uniform as possible to ensure uniform coverage of the entire material. The methods developed by the gas channel are adapted to meet these criteria. Of these, chemical vapor deposition (CVD) appears as an attractive technique for coating materials of all kinds of simple to complex shapes.

The major shortcoming of C/C composites is their low oxidation resistance. Two approaches has well been used; within it by adding inhibitors or surface deposition as thermal barrier coating.

Thermodynamic Aspect of Oxidation

The predominant products formed during the oxidation of carbon are the carbon mono-and di-oxide gas. Note that the CO (g)/CO_2 (g) ratio increases with increasing temperature but tends to decrease with an increase in the oxygen partial pressure. Indeed, this behavior is confirmed by the evolution of the free enthalpy of these two oxidation reactions with temperature. Beyond 800 ° C, carbon monoxide is the predominant product.

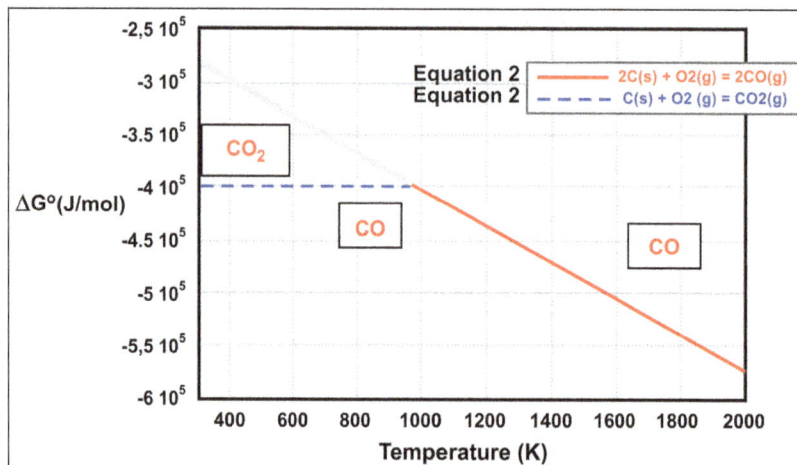

Evolution of the difference in standard molar Gibbs energy depending on the temperature.

Macroscopic Aspect of the Oxidation

The carbon gasification reaction is based on the principle of a heterogeneous solid-gas reaction governed by the following steps:

- External transfer through the boundary layer around the material of the gaseous reactants,

- Internal transfer of gaseous reactants within the porous medium,

- Adsorption of the reactants on the surface,

- Heterogeneous chemical reaction in adsorbed phase leading to the formation of products,

- Desorption of products crossing in reverse order the same steps to final diffusion through the boundary layer.

Kinetics of Oxidation

The rate of oxidation of C/C composite increases with temperature and decreases with the duration of exposure to air. Gadiou explain the control of the reaction by the oxidation reaction of oxygen on the active sites of the carbon at temperatures below about 800 °C. This results in increasing the specific surface and the surface active sites (ASA) of the material.

Up to 900 °C, it is the diffusion of oxygen through the defects of carbon, which is the limiting step. Beyond 900 °C, the mechanism is governed by the diffusion of oxygen through the boundary layer of the composite surface.

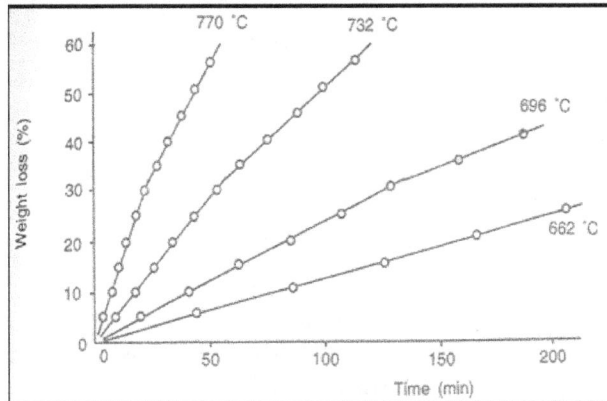

Evolution of the mass loss of C/C composites in air as a function of time for four different oxidation temperatures.

Parameters Influencing the Oxidation of C/C Composites

The oxidation resistance of C/C composites may vary significantly with the change in precursor and the process. The main reasons lie in the microstructure and microtexture of the material. The porosity of the composite is an initial response. Indeed, the open pores of the material play a major role in its behavior under air. Generally located along the fibers, they provide diffusion paths for oxygen to the interior of the composite. The interfaces between the fibers and the matrix are also to be considered. The difference in coefficient of thermal expansion (CTE) between the fiber and the matrix generates microcracks in the matrix during HTT. Thermomechanical stresses generated subsequently create larger cracks in which the oxygen can diffuse easily. This results in the destruction of the fiber/matrix bond. The microtexture of the matrix plays an important role in the oxidation resistance of the material. It appears that the graphitized carbon exhibit better resistance in air. This is explained by the fact that graphitization treatment at high temperature decreases the active sites of the carbon surface capable of reacting with oxygen. Impurities such as metal elements (Fe, Mn, Ni, Co, etc) are known as catalysts for the oxidation of C/C composites.

The second important parameter that may affect the oxidation of C/C composites is the nature of oxidant itself. Indeed, the presence of water vapor in the air can influence significantly on the reaction rate. All these factors indicate the need for PAO.

Protection of C/C Composites

Methods of Protection

The protection and the lifetime of C/C composites in an oxidizing atmosphere depend on the ability of PAO to protect the fiber reinforcement and interphase by filling microcracks by a condensed oxide phase or by consumption of gaseous oxidizing species along these microcracks.

To satisfy these requirements, various materials and means of manufacture may be envisaged However, protection can be divided into two distinct categories:

- Oxidation inhibitors, which provide protection even within the material by chemically blocking the gasification composite.

- Surface deposition, which protect the composite by limiting the access of the substrate to oxidation.

Oxidation Inhibitors

Gases are evolved from within the carbon-carbon composite during the oxidation process, mainly arising from active sites on the edges of layer planes and vacant sites formed by dislocations in the basal planes. These sites can be blocked or poisoned and the composite porosity reduced by using inhibitors based on B, P and halogen compounds which lower the reactivity of C with O_2. After impregnation of orthoboric acid into the C/C composite and subsequent processing, boron oxide (B_2O_3) is formed. An addition of 2% by weight significantly improves the oxidation resistance of the composite in dry air at 700 °C. By increasing the mass ratio of B_2O_3 the formation of a continuous layer of oxide that can heal the microcracks present in the coating when the latter is exposed to air at high temperature. However, the disadvantage of this compound is in its maximum operating temperature and atmospheric conditions. In fact, its effectiveness has been estimated at 700 °C for long-term applications (ten hours). However, this compound starts volatilize from 1000 °C, this being aggravated by the presence of moisture. At 1300 °C, B_2O_3 disappears completely after 80 hours while exposed to humid air.

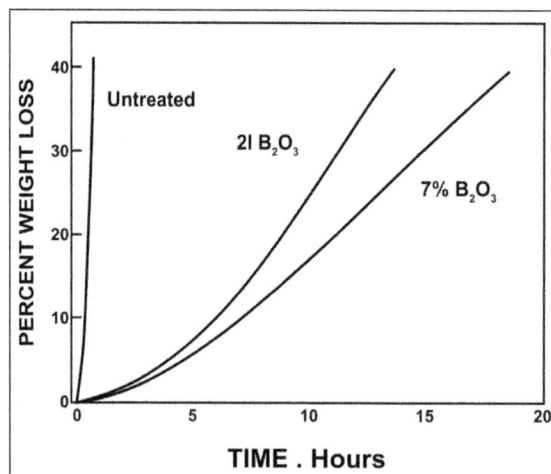

Mass loss versus time of a composite; uninhibited, inhibited 2% and 7%.

These include mixtures with metal borides: hafnium boride HfB_2 and zirconium boride ZrB_2. As these compounds are introduced in the matrix of C/C composites, during its development, this allows the formation of glassy compounds such as borate glasses with improved properties as compared with borates refractories.

Fitzer mentions the use of phosphates for protection against oxidation. $Zn_2P_2O_7$ compound can almost leave intact a C/C composite at 500 °C. These inhibitors, however, are used for protection at low temperature (below 800 °C) and do not represent a satisfactory solution for high temperature applications.

Moreover, the effectiveness of these compounds is effective only from the time the oxygen has already penetrated into the composite. The major drawback is a partial combustion of carbon in order to obtain effective protection for a user of these inhibitors.

Boron

The compound B_2O_3 is widely used for protection up to 1500 °C through poisoning sites as well as acting as a glassy sealant. Although in use it is limited to about 1000 °C due to its volatility, as a crack sealant, the working temperature can be extended to 1200 °C and to about 1400 °C if coated with an outermost layer of SiC. The glass former migrates to the outer surface, sealing cracks and voids en route. At 1525 °C, the CO liberated as a result of the B_2O_3 reacting with the carbon-carbon surface causes the partial pressure generated to disrupt the glass.

$$2B_2O_3 + 7C = B_4C + 6CO$$

Coupled with corrosion of SiC by the borate glass, the maximum temperature of use is limited to 1500 °C for short periods. TiO_2 is soluble in B_2O_3, thus increasing the viscosity and helping to prevent volatilization. Boron coatings are susceptible to attack by water at ambient temperatures, forming HBO_2.

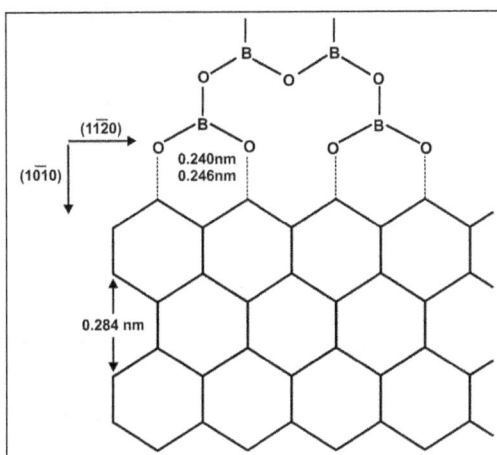

Schematic representation of bonding of $(BO_3)n$ polymer to {1010} face of graphite lattice.

Figure shows how boron can be absorbed at vacant edge sites of graphite crystallites and whilst a 3.5% B_2O_3 addition will effect a significant reduction in oxidation, a 10% addition will reduce the oxidation rate by an order of magnitude, but increase the overall weight of the composite.

Phosphorus

Organo-phosphorus compounds, when heated to 400 °C, form a phosphate residue that is tenaciously held on the carbon surface, acting as a poison, blocking active sites and reducing the rate of oxidation up to working temperatures of 850–900 °C. This temperature can be extended when phosphates are used in conjunction with other inhibitors like B_2O_3, SiO_2 and SiC. Phosphates are not as effective as borates.

Barrier Coatings

Ideally, a barrier coating should adhere well to the substrate and be free of cracks, with a coefficient of thermal expansion close to the value for the substrate. The coating should have good high temperature resistance and prevent O_2 ingress and C egress. For PAO at high temperature, surface depositions as protective coatings are preferred. Many studies have been conducted to determine the structures and compositions of materials for a viable given temperature range. Savage recommended classification of PAO in three distinct areas of temperatures:

- Low temperatures: below 1500 °C.

- The average temperatures: between 1500 and 1800 °C.

- High temperature: beyond 1800 °C.

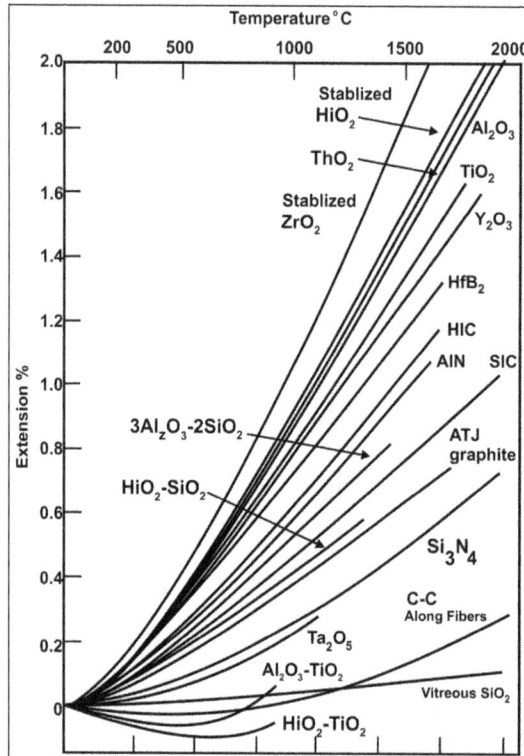

Comparison of the thermal expansion characteristics of some refractory materials.

Protections at Low Temperature

Two Basic Compounds: Carbide and Silicon Nitride

For less than 1500 °C, the silicon-based compounds are most widely reported in the literature. These materials, such as silicon carbide (SiC) or nitride (Si_3N_4), form silica precursors, which have the advantage of being a diffusion barrier against oxygen. With a CTE similar to that of a C/C composite, SiC has the advantage of being oxidized very slowly at very low temperatures (below 1000 °C).

However, this compound cannot be used alone for two major reasons. Below 1200 °C, the viscosity of the silica (SiO_2) formed by oxidation is too high for the next filling of defects and microcracks in the coating, which are preferential paths of oxygen diffusion. This phenomenon, also called "healing" can be obtained by addition of boron compounds (B, HfB_2, B_4C). Their oxidation product, boron oxide, formed within the coating has a low melting point and a lower viscosity enough to seal the microcracks.

Another drawback relates to the oxidation of silicon carbide. Depending on the oxygen partial pressure in contact with the SiC, the formation of SiO gas is preferred over that of SiO_2.

Composition Gradient Materials

To overcome these problems, the development of composition gradient materials appears a better solution. Indeed, their main advantage is their excellent adhesion to the substrate and absence of microcracks development. By gradually varying the chemical composition of the deposit, the mechanical characteristics (CTE, Young's modulus) in the material varies gradually to accommodate the properties mismatch of the substrate and the coatings. Thus, Kowbel developed two different methods for applications below 1500 °C. The first consists in the realization of a co-deposition of CVD SiC and carbon in contact with the C/C composites. This system is finally completed by an outer layer of Si_3N_4. The second technology is the use of reactive CVD by reacting SiO gas on the composite. Deposition of silicon is created and converted into Si_3N_4 by nitriding.

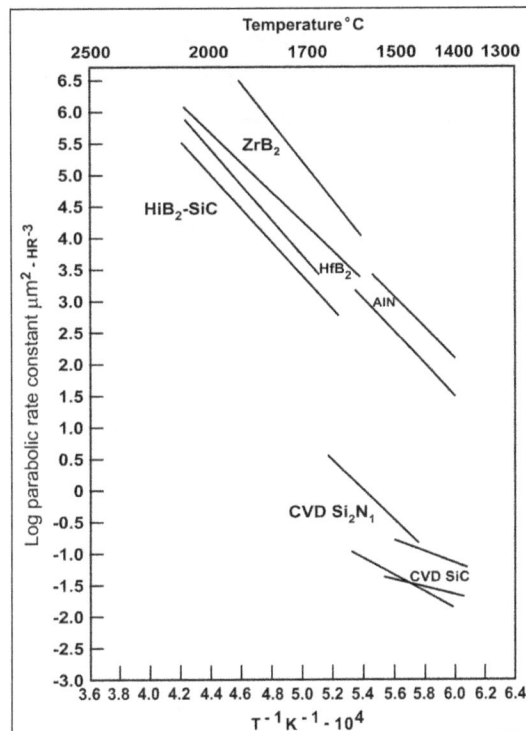

Oxidation characteristics of high temperature ceramics.

Development of Self-healing Coatings

This method, widely reported in the literature, is based on the development of self-healing coatings consist of Si-BC and interphase based on boron nitride BN.

Two layers of SiC and an intermediate layer comprising a boron compound capable of forming B_2O_3 (here: SiBC and B_4C) was used. This type of coating has two key roles for the protection of the composite. First, it consumes oxygen by oxidation of the coating and enhances its high temperature capability by propagation through the fiber reinforcement and clogging microcracks by oxide matrix as a healing phase. On the other hand, it effectively deflects cracks and prevents them from reaching the fiber reinforcement and lengthening the diffusion path of oxygen.

Protections at Intermediate Temperatures

In view of the high volatility of B_2O_3 beyond 1500 °C, this oxide can undergo carbothermal reduction leading to the formation of CO gas and B_4C. The gassing of carbon monoxide causes the appearance of bubbles in the protective undercoat causing its decoherence. We can make the same remarks for silica with the formation of gaseous silicon monoxide SiO.

Considering these observations, it appears that the SiC and B_4C cannot be used alone and a sub-layer in contact with the C/C composite is required. Carbides having a low apparent diffusivity for carbon play the role of diffusion barrier. Ultra-refractory HfC, ZrC, TiC and TaC carbides possess this property well and appear as potential candidates for the development of a sub-layer. So these carbides have interesting properties but other alternatives have been studied in the past with a particular interest in oxides. Thus Savage coupled a stack of a SiC sub-layer system with a coated layer of a refractory silicate.

Mullite $Al_6Si_2O_{13}$ has also been used as a protective outer layer of SiC. The advantage of this system lies in the coexistence of a liquid phase in the case of joining SiC + mullite. In fact, from 1600 °C, CO bubbles are formed at the cracks formed in the SiC. These are subsequently coated with the liquid phase.

However, the major drawback is the cracking of the protective layer of silica at 1650 °C due to the production of very volatile compounds such as SiO and CO formed by interfacial reaction with the SiC.

Protections to Very High Temperatures

Beyond 1800 °C, the coupling system to the SiC refractory silicates has insufficient physical and chemical properties. Other approaches have been explored involving refractory compounds capable of withstanding very high temperatures. These may include borides (ZrB_2 or HfB_2), oxides (ZrO_2, Al_2O_3, HfO_2, ThO_2, Y_2O_3), carbides (HfC, ZrC, TiC and TaC) or a combination of these materials.

A multilayer coating is the best route to impart adequate protection against oxidation and erosion of carbon-carbon composites. The outer refractory oxide coating can be ZrO_2, HfO_2, Y_2O_3 or ThO_2, with a SiO_2 glass inner layer as a sealant and oxidation barrier, an inner refractory oxide layer to provide chemical compatibility with the glass and finally, an inner refractory carbide layer of TaC, TiC, HfC or ZrC to ensure chemical compatibility with the carbon.

PAO-multilayer Approach

Kaplan et al. have patented a process in which they have stacked twenty HfC-SiC layers. The thickness of individual layers of SiC (approximately 3 microns) is greater than the thickness of individual layers of HfC (1 to 2µm) so that the CTE of the overall PAO is similar to that of SiC, the latter having an intermediate expansion coefficient between that of C/C composite and HFC. Playing on the number of interfaces present in the PAO, the oxidation resistance is improved due to the deviation at each interface cracks, preferential diffusion paths of oxygen to the C/C composite.

PAO-oxide based Approach

For very long time oxidation beyond 1800 °C, Savage recommended ultra-refractory mixed oxides: ZrO_2, HfO_2, Y_2O_3, Al_2O_3 and ThO_2. These systems need to be associated with sub-layer transition. Strife Sheenhan proposed a stack of four layers according to the following sequence -HfO_2/SiO_2/HfO_2/HfC. The silica layer, playing the role of back diffusion barrier to oxygen, effectively stops the propagation of cracks generated in the outer layer of HfO_2.

PAO-noble Metals based Approach

It is an attractive way of protecting C/C composites. With a high ductility at high temperature, materials such iridium alone or in combination with rhenium can be considered.

Iridium is most often included and has been used in many metal alloys for high temperature applications. It has the advantage of being a barrier to oxygen very effective (the permeability is lower than that of silica). Moreover, its refractoriness (melting temperature around 2400 °C) and chemical stability vis-à-vis the carbon make it a potential candidate for a high temperature barrier of oxygen. However, the oxidation resistance of iridium is strongly altered with the formation of volatile species and IrO_3 and IrO_2.

Other alloys were also tested: Zr-Pt, Hf-Ta, Pt-Si but still very inadequate protections. The review of different methods of PAO dictates to use a combination of refractory compounds. It appears to be essential for any PAO which can withstand very high temperatures. Chemically compatible multilayer assemblies (like: carbide/carbide or boride/boride) is an interesting solution.

The temperature range in which the composite must be protected influences the chemical nature of the materials constituting the PAO. Whatever the temperature range, it must meet the minimum criteria shown in the figure.

Thermomechanical and Thermochemical Behavior

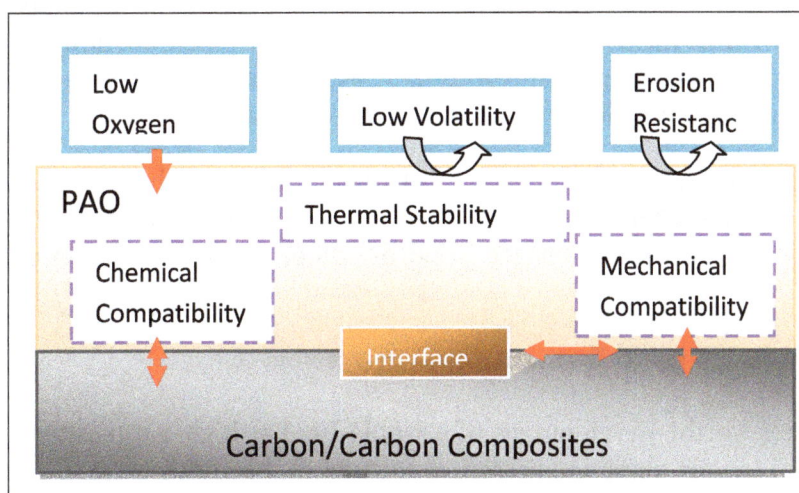

Physicochemical selection criteria of PAO for the implementation of a surface coating.

First, it is essential that the carbon from the substrate does not diffuse through the coating to prevent the carbothermal reduction of the oxides formed on the surface under the action of oxygen. In addition to being a barrier to carbon, the main function of a PAO is to act as an effective diffusion barrier to oxygen, to prevent its penetration into the material to be protected. A coating with a low permeability to oxygen will therefore constitute a protective material capable of slowing the diffusion of oxygen within the C/C composite. From this standpoint, the most effective protection are oxides such as silica. The microstructure of the coating is also taken into account; the low permeability of a surface deposit will be useless if it is cracked. The main sources of cracking can be of three different types.

First Cause: The Material Structure

The first cause may be a modification of the structure of the material during a change of temperature resulting in a change of volume element. The two most refractory oxides ZrO_2 and HfO_2 perfectly illustrate the problem of polymorphism (transition between monoclinic / tetragonal), the source of cracking 2nd case: the thermal stresses

Second Cause: The Thermal Stresses

The second probable origin of the cracking in a surface deposit or at the interface with the carbon substrate. They can appear during the development of the deposit including cooling from high temperature exposure. These induced stresses also depend on the difference of thermal expansion coefficients between the composite and the coating. To overcome this problem, a good deal between the CTE of the composite and deposit is required. In the case of too large a gap, the introduction of a sub-layer with intermediate expansion coefficient can be an interesting alternative. However, if cracks appear inevitably their deflection by a ductile material (i.e. rhenium) or anisotropic (i.e. pyrocarbon) suitably placed on the surface of the substrate or deposit effectively retard the propagation of cracks to the carbon substrate.

Third Cause: The Anisotropy of the Material

The last probable source of cracking is the anisotropy of the PAO. This problem is often solved by varying the composition of the material at the interface with the introduction of a gradient composition for continuously variable mechanical properties. The main advantage of this method is the distribution of original anisotropic stress over a larger volume.

In addition, the compatibility at the interface between carbon and the deposit can promote good adhesion through a strong chemical and mechanical bond. This may result either in the interpenetration of the materials at their interface in porosities or in-situ formation of a third phase by chemical reaction.

Analysis of Intrinsic Properties by Family

Potential candidates for the development of a viable PAO at high temperature are mostly theoretical. The state of the art possible protection showed that borides, oxides, nitrides or carbides provide good results. However, it was previously shown that the refractoriness of PAO played an important role on the temperature stability of the coating and will exceed 3000 °C. Figure shows the potential candidates according to their melting or sublimation.

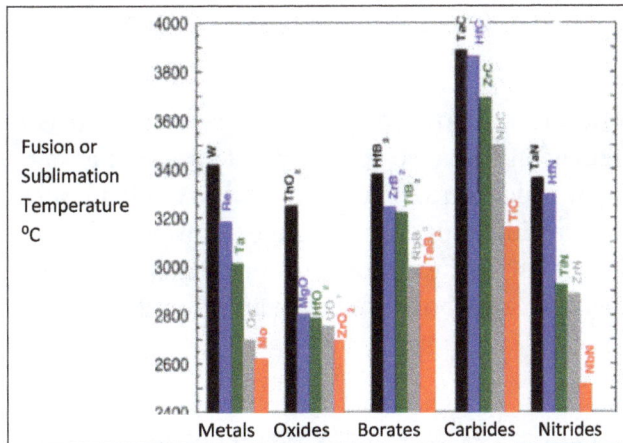

Classification by family of refractory compounds according to their melting temperature.

The most refractory ceramics are carbides of transition metals of IVa (Ti, Zr, Hf) and Va (Nb, Ta) with melting temperatures above 3000 °C tend to favor these compounds compared to borides or oxides. W and Re metals were removed due to their low resistance to oxidation. In terms of thermochemical and thermomechanical compatibility, interaction of these materials with the carbonaceous substrate must be taken into account. Between the ambient temperature to 2000 °C, the CTE of C/C composites (density ≈ 1.9) vary b/w -2 and +2 x 10^{-6}/°C.

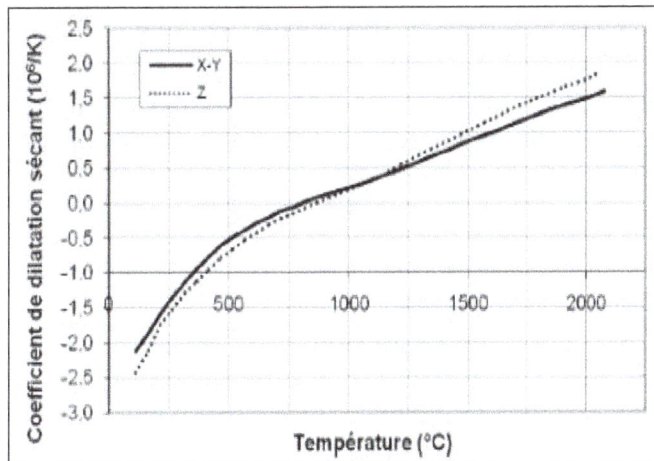

Evolution of the CTE of a C/C composite as a function of temperature.

The composite thus has the particularity to contract at room temperature. This element must be taken into account. Finally, the selected protection must also be thermally stable. A low vapor pressure of various ceramic will be an important factor in the choice of the final PAO. The analysis of all these properties is performed here by family of materials.

Oxides

Oxides are the first and largest group of ceramic materials. They possess a good chemical inertness and good high temperature properties. Apart HfO_2, they have relatively high CTE whose value reaches 10 x 10^{-6}/°C for zirconia, which is too far from that of a C/C composites. While having low vapor pressures, these compounds will expand significantly up to 2% at 2000 °C for the most refractory oxide (ThO_2, HfO_2 and ZrO_2). They therefore require the introduction of one or more sublayers for adaptation.

Another major drawback lies in their low bonding with carbon in the early filing of the PAO. This therefore results in poor adhesion. Because of their good ion-conducting property, the oxides have some permeability to oxygen which varies according to the oxide in question. Silica appears to be the best oxygen barrier over a wide temperature range. However, this oxide has a very high vapor pressure. Finally, the use of most of these oxides to protect C/C composite is subject to their reduction by the carbon substrate or carbon monoxide and therefore constitutes a major drawback.

Borides

Although boron forms stable borides with many elements such as carbon or nitrogen, only the transition metal borides are of particular interest for the protection against oxidation at high temperatures. From the viewpoint of refractoriness, the borides of transition metals of groups IVa (Ti, Zr, Hf) and Va (V, Nb, Ta) appear to be the most attractive.

Such refractory borides, due to the configuration of the boron structure, have the metallic properties such as high thermal and electrical conductivities. Chemical stability, which is directly related to the electronic structure of metal borides decreases from group IVa to the borides Va group. Thus, the most stable boride such as TiB_2, ZrB_2 and HfB_2 are of particular interest.

The melting point exceeds 3000 °C. Their CTE is lower than that of the oxides but is still far from that of a C/C composites. However, the major limitation for the use of such protection is the gradual evaporation of boron oxide formed from 1000 °C. Below this threshold temperature, a clogging phenomenon of the boron oxide is ensured due to the high viscosity of the oxide.

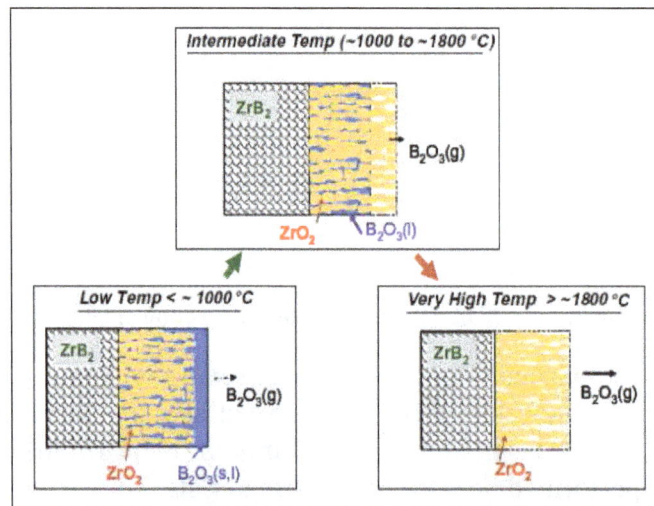

Evolution of the oxidation of ZrB_2 in air as a function of temperature.

Above 1800 °C applications, when used with HfO_2, the boron oxide is completely volatilized leaving columnar and porous surface structure of hafnium HfO_2.

In the case of reusable objects, another obstacle to the use of refractory borides is their ability to form intermediate phases during cooling which is a source of unstable interface in contact with the composite.

Furthermore, the presence of a eutectic at a temperature of 1800 °C to 87 atomic% of hafnium is an additional problem with the use of refractory borides.

Nitrides

Among all nitrides, refractory compounds of Group IVa (HfN , ZrN and TiN) are those of greatest interest from the viewpoint of refractoriness. They have melting temperatures broadly equivalent to those of refractory borides. The hafnium nitride is the only compound whose CTE approaches that of borides. However, HfN having over-stoichiometry nitrogen atoms ($HfN_{1,12}$) can create instability in contact with the composite or another refractory element in the context of a multilayer.

Carbides

The carbide ceramics are the family of most chemically compatible with the C/C composites. As the borides, carbides of Groups IVa and Va are of greatest interest from the viewpoint of refractoriness. The melting point of hafnium and tantalum carbides is around 4000 °C. The CTE of the refractory carbides is generally similar to that of borides. The carbides of Group IVa (Ti, Zr, Hf) appear to be satisfactory compounds, keeping in view of the various compatibility criteria. High temperature stability of carbides of Va group with low vapor pressures in excess of 2500 °C, these carbides have the distinction of being regarded as interstitial carbides. Indeed, these compounds admit a large area of non-stoichiometry (M_xC_{1-x}).

Sub-stoichiometric carbides have a better adhesion to the carbon substrate. This non-stoichiometry is accompanied by a change in physical properties e.g. lattice parameters, micro hardness etc.

Carbo

The formation of carbonitrides of refractory metals (i.e. those of group IVa) is possible (the nitride and carbide of the Group IVa metals are completely miscible and crystallized in the same structure as that of NaCl). They would present a better resistance to oxidation than the corresponding carbide by filling gaps of carbide M_xC_{1-x} by nitrogen atoms. Despite their interesting mechanical properties (TiCN use in the field of cutting tools) the use of these materials as PAO has not been considered.

Combination of Properties: Carbide/Carbide Multilayer System

These carbides are therefore preferred compounds for the protection of composite C/C. Hafnium carbide (HFC) along with tantalum carbide (TaC) has is a potential candidate particularly due to their thermal stability at very high temperatures. Moreover, the oxidation of this compound from hafnia HfO_2, a refractory oxide with significant oxygen permeability. Therefore, it is necessary to carefully analyze the properties of oxidation resistance and include, if necessary, another carbide which enhances protection of the composite.

5

Nanocomposites

A multiphase solid material in which one of the phases has one, two or three dimensions of less than 100 nanometers is called nanocomposite. There are different types of nanocomposites such as ceramic-matrix nanocomposites, metal-matrix nanocomposites and polymer-matrix nanocomposites. This chapter discusses in detail these types of nanocomposites.

Nanocomposites are the materials of twenty-first century having an annual growth rate of 25% due to their multifunctional capabilities. With unique design possibilities and properties, they attract the attention of researchers worldwide. Due to the possibility of combining desired properties, nanocomposites are expanding their potentials in aerospace applications and in future space missions. Selection of the constituents of nanocomposites (matrix and nanofillers) leads to the improvement of certain desired properties. For aerospace applications, mechanical, thermal, electrical, chemical and biodegradable properties are of interest. Chemical property like resistance or passiveness to corrosion is of prime importance. Apart from low weight requirements, aerospace structures pose requirement of mechanical properties for design like strength, toughness, fatigue life, impact resistance and scratch resistance. Aircrafts flying at high altitudes require properties like low solar absorption, radiation resistance, high thermal emissivity and electrical conductivity. Due to increasing use of nanocomposites, handling their disposal after the service life is of great concern. Thus, biodegradable property is also being preferred apart from their functional properties. Being environmentally friendly, few nanocomposites have opened the possibilities of clean technologies. Proper choice of nanofillers and matrix is a key in achieving the desired multifunctional properties that are required by many aerospace applications.

Nanofillers with low volume of 1–5% can enhance the properties of composite materials that are comparable to those of conventional microfillers with 15–40% volume. Nanofillers have exceptional properties since they have basic structure at crystal level that is defect free. Nanomaterials can be classified as per the dimensions. Iso-dimensional (three-dimensional) nanoparticles have the three dimensions in nanoscale like silica, metal particles and semiconductor particles. Nanotubes or whiskers are second kind of two-dimensional nanoparticles that have two dimensions in nanometer scale (<100 nm) with the third dimension forming an elongated structure (aspect ratio more than 100). Fillers like sheets are one-dimensional with one dimension like the thickness is in nanometer range (1 nm) and the other two dimensions have an aspect ratio >25. Each type of nanofiller has its own advantages, disadvantages and unique properties.

Varieties of nanocomposites have been evolved depending upon the type of matrix used for which they are widely classified into three categories (as in case of microcomposites) as polymer matrix nanocomposites (PMNCs), ceramic matrix nanocomposites (CMNCs) and metal matrix nanocomposites.

Ceramic-matrix Nanocomposites

Ceramic matrix nanocomposites combine the properties of ceramics like wear resistance, high thermal stability and chemical stability with the properties of nanofillers.

In this composite group, a ceramic such as a chemical compound from a group of nitrides, oxides, silicides and borides occupy the major part of the volume. In most ceramic-matrix nanocomposites, a metal is the second component. In an ideal scenario, both the ceramic and the metallic components are dispersed evenly to elicit specific nanoscopic properties.

The binary phase diagram of the mixture should be considered in designing ceramic-metal nanocomposites and measures have to be taken to avoid a chemical reaction between both components. The last point mainly is of importance for the metallic component that may easily react with the ceramic and thereby lose its metallic character. This is not an easily obeyed constraint because the preparation of the ceramic component generally requires high process temperatures. The safest measure thus is to carefully choose immiscible metal and ceramic phases. A good example of such a combination is represented by the ceramic-metal composite of TiO_2 and Cu, the mixtures of which were found immiscible over large areas in the Gibbs' triangle of Cu-O-Ti.

The concept of ceramic-matrix nanocomposites was also applied to thin films that are solid layers of a few nm to some tens of μm thickness deposited upon an underlying substrate and that play an important role in the functionalization of technical surfaces. Gas flow sputtering by the hollow cathode technique turned out as a rather effective technique for the preparation of nanocomposite layers. The process operates as a vacuum-based deposition technique and is associated with high deposition rates up to some μm/s and the growth of nanoparticles in the gas phase. Nanocomposite layers in the ceramics range of composition were prepared from TiO_2 and Cu by the hollow cathode technique that showed a high mechanical hardness, small coefficients of friction and a high resistance to corrosion.

Alumina/MWNT for EM Shielding Applications

Alumina is typically used in structural applications but is an electrical insulator for which it cannot be used in aircrafts. CNTs are seen to impart toughness to the ceramic matrix with improved properties compared to the conventional carbon and SiC fibers, which just imparts toughness. MWNTs show multichannel, quasi-ballistic conducting behavior due to the participation of multiple walls in the electrical transport. Their large diameter results in small band gap for semiconductor activity. High current density and conductivity (10^7 A/cm and 1.85×10^3 S/cm, respectively) of MWNTs render them electrically conductive always. Loading of MWNTs in alumina results in improved structural integrity and electrical conductivity. Thus, it is worth to present the preparation method and alternating current (AC) and DC conductivity studies of alumina/MWNT nanocomposites developed by Ahmad and Pan. MWNTs were mixed ultrasonically with alumina in ethanol followed by ball milling for 24 h. The mixture was dried and transferred in cylindrical graphite mold to be plasma sintered at 1400 °C in vacuum under a pressure of 50 MPa.

A SEM of MWNT/alumina nanocomposite (fractured surface) with 3 wt% MWNT. A well-dispersed intertwining network formed by MWNTs at the grain boundaries can be seen. Conduction via tube–tube within the bundle and interlinked network of ropes results in the overall conduction in such nanocomposite. A standard four-point probe method gave the DC conductivity as shown in figure. Insulator conductor transition was exhibited between 0.4 and 0.5 wt% of MWNTs. Eight order increase in conductivity was observed at 0.5% MWNT after which a lesser rise was observed. The power law followed by the conductivity is given by,

$$\sigma_{dc} = \sigma_c (P_{MWNT} - P_c)^t, \text{ for } P_{MWNT} > P_c,$$

where σ_{dc} and σ_c are the DC conductivities of the nanocomposite and the filler, respectively. P_{MWNT} and P_c are the weight fraction and percolation threshold, respectively.

SEM showing a fractured surface of MWNT/alumina nanocomposite with uniform distribution of MWNTs.

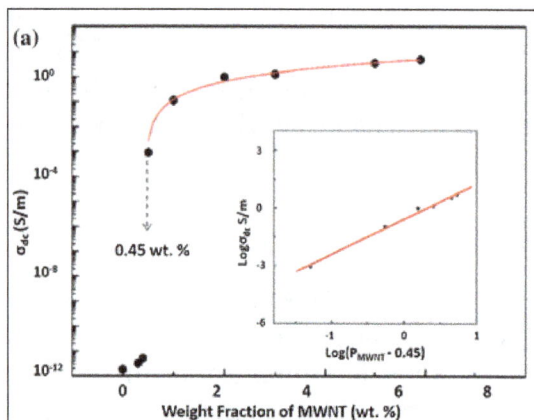

(a) DC conductivity of the MWNT/alumina nanocomposites as a function of MWNT contents. Inset shows DC conductivity with (P_{MWNT}–P_c). (b) Frequency-dependent conductivity of the MWNT/alumina nanocomposites.

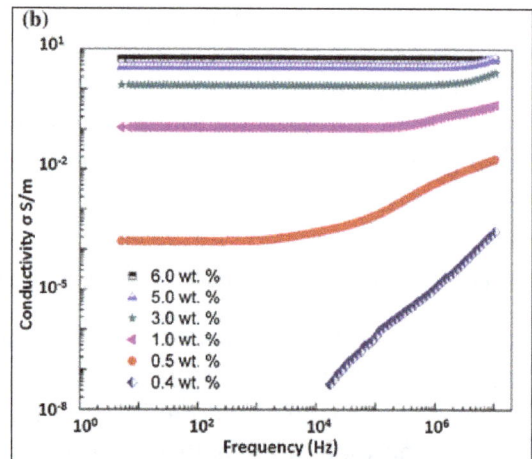

The fit in figure has the $t = 1.82$ (represents the dimensionality of the system) and $P_c = 0.45$ wt%. A smaller amount of t is contributed by the hopping between weakly connected parts of the network. For $t < 2$, thermally induced hopping transport dominates the conduction. Increase in wt% of MWNTs enhances the conductive path by increasing the inter-nanotube connections. Tunneling resistance is caused by inter-nanotube connections. The conductivity of the nanocomposites is also seen to increase with temperature. The carrier mobility and thermally generated carrier mobility increase with temperature causing semiconductor-like effect. Fluctuation-induced tunneling model is used to describe this effect as a change of barrier height by fluctuation of thermal voltage causing the tunneling of the electrons.

$$\sigma_{dc} \propto \exp\left[\frac{-T_1}{T + T_0}\right],$$

where $T_1 = wA\varepsilon_0^2 / 8\pi k_B$, $T_0 = 2T_1 / \pi w\chi$, T is the absolute temperature, k_B is the Boltzman's constant, $\chi = (2m_c V_0/h^2)^{1/2}$, $e_0 = 4V_0/ew$, m_c is the electron mass, e is the charge, V_0 is the potential barrier height, w is the insulating layer width and A_J is the area of capacitance formed at the junction.

AC conductivity studies were done using an impedance analyzer at frequency ranging from 5 to 15 MHz at room temperature. The conductivity was initially constant and increased at higher frequencies for samples just above the percolation threshold (0.4 wt%). For higher MWCNT percentage, the conductivity becomes frequency independent. Below percolation threshold, the conductivity increases linearly with the slope of curve closer to unity indicating highly resistive material. The curve obeys the expression of dielectric conductivity $\sigma = 2\Pi fee_0$, where e_i is the imaginary part of permittivity and e_0 is the vacuum permittivity. In addition to the electrical properties, they possess excellent resistance to chemical, corrosion, extreme radiation and ultraviolet exposure generally encountered by aerospace structures. With these properties, MWNT alumina nanocomposites are suitable as electrostatic dissipation materials for aerospace structures.

Alumina/MWNT for SHM Applications

Apart from applications like energy, transportation, defense, sporting goods and civil infrastructure, CMNCs are being researched for automotive and aerospace applications. Automotive and aerospace applications demand SHM capabilities for safety of life and infrastructure. Inam et al. presented the SHM capability in ceramic/carbon nanocomposite. SWNTs were dispersed in DMF by ultrasonication for 30 min followed by hand mixing of alumina nanopowder (particle size <50 nm). Ball milling of the mixture was done for 7 h followed by drying in rotary dryer for 12 h (60 °C) and in vacuum over for 48 h (100 °C). The mixture was ground and sieved at 250 mesh followed by drying in vacuum oven for 48 h (100 °C). Nanocomposites were prepared by spark plasma sintering from which 18 × 3 × 3 mm bars were prepared for various tests.

The fractured microstructure of alumina/CNT nanocomposite obtained from SEM. Homogeneously dispersed CNTs in various isolated location and a CNT mesh in central fractured location indicated uniform dispersion obtained by milling and drying process. Vickers indentations were performed with 500 g load and 10 s of indentation time. Five equally spaced indents were produced with parallel measurement of bulk electrical conductivity at room temperature using a four-point method. A drastic change in electrical conductivity was seen in alumina/CNT nanocomposite with number of indents. CNT and graphene-based nanocomposites showed higher sensitivity toward indentation damage. Also superposed are the results obtained from alumina/CB and alumina/graphene nanoplatelets. The indentation causes microcracks attributing to the fracture of fibers that result in reduced conductivity. Higher fracture toughness was also seen in these nanocomposites. Thus, with improved fracture toughness and safety provided by SHM capability, alumina/CNT nanocomposites can be used as structural components of aircraft, spacecraft and reentry vehicle operating at higher temperature.

Metal-matrix Nanocomposites

The other name for metal matrix nanocomposites is reinforced metal matrix composites. This composite type can be classified as non-continuous and continuous reinforced materials. Carbon nanotube metal matrix composites is an important nano composite that takes advantage of the high electrical conductivity and tensile strength of carbon nanotube materials. The new research areas on metal matrix nanocomposites are carbon nitride metal matrix composites and boron nitride reinforced metal matrix composites.

The energetic nanocomposite is another kind of nanocomposite normally as a hybrid sol–gel with a silica base, which, that when combined with nanoscale aluminum powder and metal oxides forms superthermite materials.

A recent study, comparing the mechanical properties (Young's modulus, compressive yield strength, flexural modulus and flexural yield strength) of single-and multi-walled reinforced polymeric (polypropylene fumarate—PPF) nanocomposites to tungsten disulfide nanotubes reinforced PPF nanocomposites suggest that tungsten disulfide nanotubes reinforced PPF nanocomposites possess significantly higher mechanical properties and tungsten disulfide nanotubes are better reinforcing agents than carbon nanotubes. Increases in the mechanical properties can be attributed to a uniform dispersion of inorganic nanotubes in the polymer matrix (compared to carbon nanotubes that exist as micron sized aggregates) and increased crosslinking density of the polymer in the presence of tungsten disulfide nanotubes (increase in crosslinking density leads to an increase in the mechanical properties). These results suggest that inorganic nanomaterials, in general, may be better reinforcing agents compared to carbon nanotubes.

Another kind of nanocomposite is the energetic nanocomposite, generally as a hybrid sol–gel with a silica base, which, when combined with metal oxides and nano-scale aluminum powder, can form *superthermite* materials.

Synthesis, Processing and Properties of Metal Matrix Nanocomposites

The greatest challenges facing the development of MMNCs for wide application are the cost of nanoscale reinforcements and the cost and complexity of synthesis and processing of nanocomposites using current methods. As with conventional metal matrix composites with micron-scale reinforcements, the mechanical properties of a MMNC are strongly dependent on the properties of reinforcements, distribution, and volume fraction of the reinforcement, as well as the interfacial strength between the reinforcement and the matrix. Due to their high surface area, nanosize powders and nanotubes will naturally tend to agglomerate to reduce their overall surface energy, making it difficult to obtain a uniform dispersion by most conventional processing methods. In addition, due to their high surface area and surface dominant characteristics, these materials may also be highly reactive in metal matrices. For example, in Al/CNT composites there are concerns that brittle aluminum carbide phases could form during processing, impairing the mechanical properties and electrical conductivity of the nanocomposite. Because of these concerns, processing methods are being developed to produce MMNCs with uniform dispersions of nanomaterials and little deleterious interfacial reactions.

The methods that have been used to synthesize metal matrix nanocomposites include powder metallurgy, deformation processing, vapor phase processing, and in some cases solidification processing. Powder metallurgy involves the preparation of blends of powders of metal and reinforcements, followed by consolidation and sintering of the mixtures of powders to form the part. Deformation processing involves subjecting a metal to high rates of deformation to create nanostructured grains in a metal matrix. Vapor phase processing methods such as chemical vapor deposition (CVD) can be used to deposit thin films creating dispersed multiphase microstructures, multilayered microstructures, or homogeneous nanostructured coatings. Each of these methods can create very desirable microstructures, however they are expensive and difficult to scale up to manufacture large and complex shapes in bulk.

Of the processing methods available for synthesis and processing of MMNCs, the least expensive method for production of materials in bulk is solidification processing. There are various avenues by which researchers have created nanostructures and nanocomposite materials using solidification and these can be divided into three categories:

1. Rapid solidification,

2. Mixing of nanosize reinforcements in the liquid followed by solidification,

3. Infiltration of liquid into a preform of reinforcement followed by solidification.

Rapid solidification (implying solidification rates of up to 10^4–10^7 °C/s) through methods such as melt spinning (a liquid metal stream is impinged on a spinning copper drum), or spray atomization (a superheated liquid metal is atomized with gas jets and impinged on a substrate) can lead to nanosize grains as well as amorphous metals from which nanosize reinforcements can be precipitated in the amorphous matrix during heating to form nanocomposites. Mixing techniques involve adding particulate reinforcements and mechanically dispersing them in the matrix. Mixing methods that have been applied to synthesize MMNCs include stir mixing, where a high temperature impeller is used to stir a melt that contains reinforcements, creating a vortex in the melt, and ultrasonic mixing, where an ultrasonic horn is used to create cavitation in the melt that disperses the particulate reinforcements by a gas streaming effect that occurs through the collapse of bubbles within the melt. Infiltration techniques entail infiltrating a preform or partial matrix containing the reinforcements with a liquid metal. The preform consists of particles formed in a particular shape with some binding agent, and can be composed of the additives and binding agent alone or with some portion of the matrix added as a partial filler. Infiltration methods that have been used include ultrahigh pressure, where the pressure used to infiltrate a high-density preform of nanoparticles is in excess of 1 Gpa, and pressureless infiltration, where a block of metal is melted on top of a lower density preform of nanoparticles and allowed to seep into the preform.

In figure illustrates examples of the different microstructures of Aluminum Alloy-Al_2O_3 nanoparticle composites that have been produced by different processing methods. Figure shows a microstructure exhibited by a cast nanocomposite synthesized at UW–Milwaukee by the authors. This MMNC was made using a unique casting method combining the use of stir mixing, and ultrasonic mixing, with a wetting agent added to the molten alloy to incorporate nanoparticles in a metal matrix. This process resulted in the incorporation of nanoparticles within microscale grains of aluminum, and formed a bimodal microstructure. Ceramic nanoparticles may be uniformly dispersed in metal matrices to increase the tensile strength and wear resistance, using methods such as ultrasonic cavitation of the melt to further disperse the particles. The influence of nanoscale reinforcements on formation of solidification microstructure including their influence on nucleation, growth, particle pushing, solute redistribution, heat and fluid flow, however, will have to be understood to reproducibly create desired structures. While there is some understanding of the influence of micron size particles on the formation of solidification microstructures, the influence of nanoscale reinforcements on each of the constituents of solidification structure formation need to be studied using both theoretical and experimental research. Figure shows a transmission electron microscope (TEM) micrograph of the microstructure obtained by ball milling pure metals and nanopowders, followed by hot pressing/sintering to form a nanocomposite. The matrix aluminum and the Al_2O_3 powders are nanosize in this TEM micrograph and the reinforcement phases are mainly restricted to the grain boundaries, which will likely have a grain boundary pinning effect. Since a much higher percentage of Al_2O_3 has been added using this process, the resulting sample exhibited a substantial improvement in its wear properties.

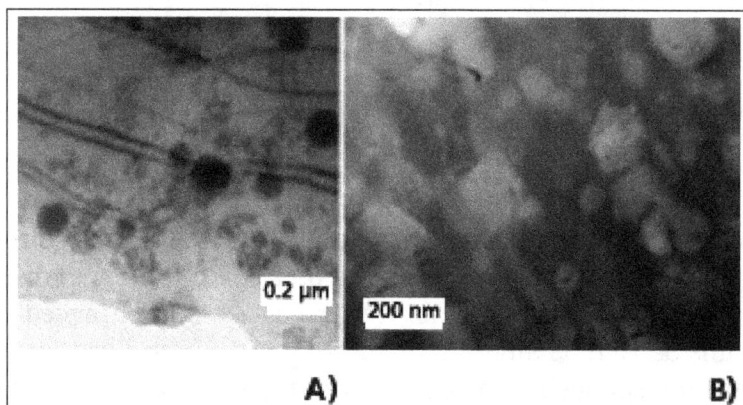

TEM of Aluminum Alloy-Al_2O_3 nanocomposites produced by liquid and solid based methods. A) Stir cast A206-2 vol % Al_2O_3 (47 nm) nanocomposite produced by the authors at UW–Milwaukee. B) Powder metallurgy based Aluminum alloy-15 vol% Al_2O_3.

Great improvements can be achieved in specific material properties by adding only a small percentage of a dispersed nano-phase as reinforcement. Table presents selected studies on MMNCs, the processing techniques used and their respective properties. When the reinforcement in a metal matrix is brought down from micron-scale to nano-scale, the mechanical properties are often substantially improved over what could be achieved using micron-scale reinforcements. This is possibly due to the exceptional properties of the individual nano-phase reinforcements themselves, smaller means free path between neighboring nanoparticles and the greater constraint provided by the higher surface area of nano-phase reinforcements. Nano-phase reinforcements like CNT and SiC have much higher strengths than similar micron-scale reinforcements. In some cases, the nanoscale reinforcement leads to property changes in the matrix itself. For instance, nanoscale reinforcements can lead to nanosize grains in the matrix, which will increase the strength of the matrix. Due to their size, properties of nanomaterials are dominated by their surface characteristics, rather than their bulk properties, which is the case in micronscale reinforcements. The potentially unique interfaces between nanosized reinforcements and the matrix can lead to even greater improvements in the mechanical properties due to the strong interface between the reinforcement phase and the matrix, and through secondary strengthening effects such as dislocation strengthening. Figure shows that as the particle size of Al_2O_3 goes from micron to nanosize, there is significant decrease in the friction coefficient and wear rate of aluminum composites. In addition, the incorporation of only 10 volume percent of 50 nm sized Al_2O_3 particles to the aluminum alloy matrix resulted in an increase of the yield strength to 515 MPa. This is 15 times stronger than the base alloy, 6 times stronger than the base alloy with 46 volume percentage of 29 micron size Al_2O_3 and over 1.5 times stronger than AISI 304 stainless steel.

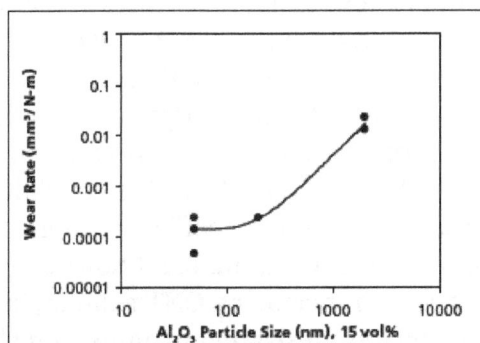

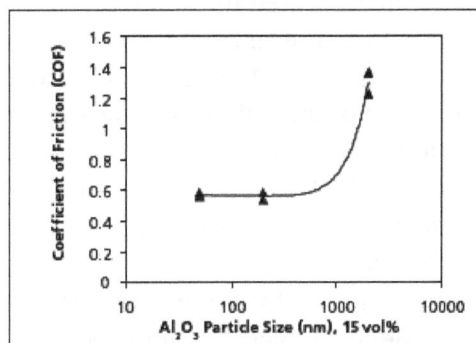

Effect of particle size on wear rate and coefficient of friction of Al-15vol% Al_2O_3 metal matrix composites. Both the wear rate and coefficient of friction are dramatically reduced when the particle size is reduced below 1 μm.

Table: Reagents for non-aqueous titration.

Team	Process	Properties and Comments
Takagi et al.	Rapidly solidified nanocrystalline Al alloys and Al/SiC nanocomposites (Matrix: Al-Ni-Y-Co, Al-Si-Ni-Ce, and Al-Fe-Ti-Me, where Me: Cr, Mo, V, Zr). SiC size: #3000 to #8000.	High hardness, strength and excellent wear resistance to 473 K.
ElEskandarany	High-energy ball milling of Al-SiC (2–10%) nanocomposite and consolidation using plasma activated sintering.	Fully dense bulk nanocomposites with nanocrystalline structure and uniform SiC distribution. No reaction products (Al_4C_3) or Si at the interface. Hardness and mechanical strength characterized.
Hong et al.	Deformation processing (drawing) of Cu/Nb filament nanocomposites.	At drawing strain above 10, a limiting thickness of Nb filaments of 10 nm was obtained (further deformation caused filament rupture rather than thinning). The ductility was independent of the Nb content.
Islamgaliev et al.	Cu-0.5wt% Al_2O_3 nanocomposites produced by high-pressure torsion technique.	High tensile strength (680 MPa) and microhardness (2300 MPa), as well as high thermal stability, creep strength and electrical conductivity were obtained.
Fekel & Mordike	Mg strengthened by 30 nm-SiC particles. Nano-SiC formed by CO_2 laser-induced reaction of Si and acetylene. Micronsize Mg powder formed by gas atomization of Mg melt with Ar gas. Composite formed by hot milling followed by hot extrusion.	Tensile Strength doubled as compared to unreinforced Mg. At room temperature increase in the UTS was around 1.5 times. Significant improvement in the hardness. Milled composites exhibited lowest creep rate.
Dong et al.	CNT/Cu composite. CNT formed by thermal decomposition of acetylene. CNT mixed with Cu powder using ball mill. Ball-powder weight ratio 6:1. Pressed at 350 MPa for 5 min. Sintered for 2 h at 850 °C.	Friction coefficient reduced with increasing CNT fraction in Cu. Less wear loss with increasing CNT content. Less wear loss in Cu/CNT than Cu/C as load increased.

Recently, an aluminum alloy specimen reinforced with CNTs was synthesized using pressureless infiltration. CNTs were mixed together with aluminum and magnesium powders using ball milling and then pressed into preforms that were subsequently infiltrated by melts of the aluminum alloy matrix at 800 °C in a nitrogen atmosphere. The CNTs were observed to be well-dispersed and embedded in the matrix. Further experiments showed that up to 20 volume percent of nanotubes could be incorporated in the matrix of aluminum alloys using this process. The wear data also suggests that the presence of CNTs in the matrix can reduce the direct contact between the aluminum matrix and the steel pin and thereby decrease the friction coefficient due to the presence of carbon nanotubes. By reducing the friction coefficient, the energy loss experienced by components in frictional contact will be reduced, improving efficiency of mechanical systems. the incorporation of CNTs having relatively short tube lengths may allow them to slide and roll between the mating surfaces and result in a decrease of the friction coefficient. Ploughing wear appeared to be the dominant wear mechanism under the dry sliding condition. The depth of the wear grooves caused by ploughing wear decreased with increasing CNT content, suggesting that the strengthening effects of nanotubes increased the wear resistance of aluminum alloy-CNT nanocomposite.

Though nanocomposite materials exhibit ultrahigh-strength, there is often a trade-off that results in decreased ductility. This may be attributed to currently used processing methods that result in the formation of voids and defects, as well as to the inability of nanostructured grains or additives to sustain a high rate of strain hardening. These shortcomings, among other microstructural effects, lead to instabilities upon plastic deformation. One of the methods that has been used to overcome the lack of ductility in nanostructured materials is to incorporate nanosize dispersoids in a bimodal or trimodal microstructure. In the case of nanostructured grains, the presence of hard precipitates or nanoparticles in a metal matrix may act to initiate, drag and pin dislocations, reducing dynamic recovery, and thus resulting in a high strain-hardening rate that, in turn, produces larger uniform strains and higher strengths in the MMNC, along with higher ductility.

The potential for achieving much higher strengths in combination with acceptable ductility in MMNCs has been demonstrated by E. Lavernia and coworkers. A mixture of a 5083 aluminum alloy powder and micron-sized boron carbide particles that had been coated with a powdered, nanostructured aluminum alloy were milled at cryogenic temperatures (cryomilled). This cryomilled powder was then mixed with regular 5083 alloy and consolidated and extruded to form nanocomposites with almost double the strength of the monolithic 5083 alloy. Cigar-shaped nanocomposite volumes of material that contained micron-sized B4C reinforcements were observed to be dispersed in nanograin sized aluminum matrix regions that were embedded in micro-grain sized 5083 aluminum matrix having no B_4C reinforcements. This combined bimodal and trimodal microstructure is presumed to be the source of the exceptional combination of strength and ductility in aluminum matrix nanocomposites that had micron size B_4C particles used as reinforcements. Apparently, the micron grain-sized aluminum matrix that did not contain any reinforcement, enabled dislocation motion and improved ductility. This type of combined bimodal and trimodal microstructure in MMNCs will need to be synthesized using less expensive solidification processing routes that can enable processing of large components having complex shapes. By exploring lower cost and more versatile methods to manufacture metal matrix nanocomposites with improved ductility, these materials are expected to become commercially viable for a variety of applications, particularly where weight savings is essential.

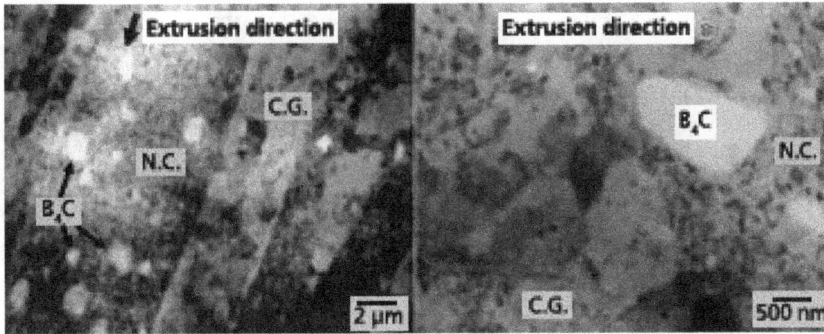

TEM micrographs showing alternate stacking of course grain size Al and nanostructured Al regions. The B_4C particles are uniformly distributed within the nanostructured aluminum exclusively.

Current and Future Applications

Metal matrix composites with micron-size reinforcements have been used with outstanding success in the automotive and aerospace industries, as well as in small engines and electronic packaging applications. In the case of metal matrix nanocomposites, incorporation of as little as one volume percentage of nanosize ceramics has leads to a much greater increase in the strength of aluminum and magnesium base composites than was achieved at much higher loading levels of micron-sized additions. Such potential improvements have great implications for the automotive and aerospace, and, in particular, the defense industries due to the drastic weight savings and exceptional properties that can be achieved. Potential aerospace applications may include ventral fins for aircrafts, as well as fan exit guide vanes for commercial airline jet engines. Both components require high stiffness and strength, low weight as well as resistance to erosion from rain, airborne particulates and hail. Components used in the automotive industry where bulk nanocomposites would likely be valuable include brake system components, which require high wear resistance, and thermal conductivity, intake and exhaust valves, which require high creep resistance and resistance to sliding wear, as well as piston liners, which require high wear resistance, good thermal conductivity and low coefficient of thermal expansion. In addition, exceptionally high thermal conductivities, possible in selected nanocomposites, will find applications in thermal management applications in computers. Metal matrix nanocomposites can be designed to exhibit high thermal conductivity, low density, and matching coefficient of thermal expansion with ceramic substrates and semiconductors, making them ideal candidates for such applications.

Polymer-matrix Nanocomposites

The addition of nanoparticulates to a polymer matrix improves its performance by using the properties and nature of the nanoscale filler. When the nanoscale filler properties are better or different from the matrix reinforcement and when the filler is dispersed well, this strategy works perfectly.

The enhancement in mechanical properties may not be restricted to strength or stiffness. By adding nanofillers, time-dependent properties can be improved. On the other hand, the improved properties of high-performance nanocomposites is because of the high surface area of fillers and the high aspect ratio as nanoparticulates have increased surface area to volume ratios when there is a good dispersion.

Nanoparticles such as graphene, carbon nanotubes, molybdenum disulfide and tungsten disulfide are being used as reinforcing agents to fabricate mechanically strong biodegradable polymeric nanocomposites for bone tissue engineering applications. The addition of these nanoparticles in the polymer matrix at low concentrations (~0.2 weight %) cause significant improvements in the compressive and flexural mechanical properties of polymeric nanocomposites. Potentially, these nanocomposites may be used as a novel, mechanically strong, light weight composite as bone implants. The results suggest that mechanical reinforcement is dependent on the nanostructure morphology, defects, dispersion of nanomaterials in the polymer matrix, and the cross-linking density of the polymer. In general, two-dimensional nanostructures can reinforce the polymer better than one-dimensional nanostructures, and inorganic nanomaterials are better reinforcing agents than carbon based nanomaterials. In addition to mechanical properties, polymer nanocomposites based on carbon nanotubes or graphene have been used to enhance a wide range of properties, giving rise to functional materials for a wide range of high added value applications in fields such as energy conversion and storage, sensing and biomedical tissue engineering. For example, multi-walled carbon nanotubes based polymer nanocomposites have been used for the enhancement of the electrical conductivity.

Nanoscale dispersion of filler or controlled nanostructures in the composite can introduce new physical properties and novel behaviors that are absent in the unfilled matrices. This effectively changes the nature of the original matrix (such composite materials can be better described by the term *genuine nanocomposites* or *hybrids*). Some examples of such new properties are fire resistance or flame retardancy, and accelerated biodegradability.

A range of polymeric nanocomposites are used for biomedical applications such as tissue engineering, drug delivery, cellular therapies. Due to unique interactions between polymer and nanoparticles, a range of property combinations can be engineered to mimic native tissue structure and properties. A range of natural and synthetic polymers are used to design polymeric nanocomposites for biomedical applications including starch, cellulose, alginate, chitosan, collagen, gelatin, and fibrin, poly (vinyl alcohol) (PVA), poly (ethylene glycol) (PEG), poly (caprolactone) (PCL), poly (lactic-co-glycolic acid) (PLGA), and poly (glycerol sebacate) (PGS). A range of nanoparticles including ceramic, polymeric, metal oxide and carbon-based nanomaterials are incorporated within polymeric network to obtain desired property combinations.

Structural Applications

In some applications, apart from being lightweight, high performance is of prime importance in aerospace structures e.g., equipment enclosures, aircraft interiors, coatings, cockpit, crew gear, heat shrinkage tubing, space durable mirrors, housings, shrouds, nozzles and solar array substrates. Composite materials offer chemical stability and fire resistance apart from the advantage of low operating cost due to their lightweight. Three major drawbacks limit the applicability of composites in most aerospace structures. Firstly, a higher electrical resistance leads to the limitation for applications like electromagnetic shielding, circuits, antennas and lightning strike protection. Secondly, lower thermal conductivity of composites increases the load on deicing systems based on electrical heaters. Thirdly, composites are less resistant to impact and suffer from moisture absorption, environmental degradation and aging with time. Depending upon the matrix and filler used, many solutions have been offered by nanocomposites to overcome these limitations.

Polyimide/Clay

Clay or layered silicates (LS) are the nanofillers that intercalate within the polymer to form layered nanocomposites. These are readily available at low cost and yield predictable stiffening in the nanocomposites. The ability to tune the surface chemistry via ion exchange reactions with organic and inorganic cations yields strong bonding between LS and polymer matrix. Due to the easy availability and well-known intercalation chemistry, clays are widely used in PMNC. Compared to microfiber composites, LS-based nanocomposites show improvements in moduli, strength, heat resistance and biodegradability along with decreased flammability and gas permeability. These improvements make LS-based nanocomposites attractive for aerospace applications. Polyimide/clay nanocomposite is prepared by sol–gel process that involves embedding monomers and organic molecules on sol–gel matrices. Chemical bonds formed by the introduction of organic groups lead to the formation of sol–gel matrix within the polymer, simultaneously forming organic/inorganic networks.

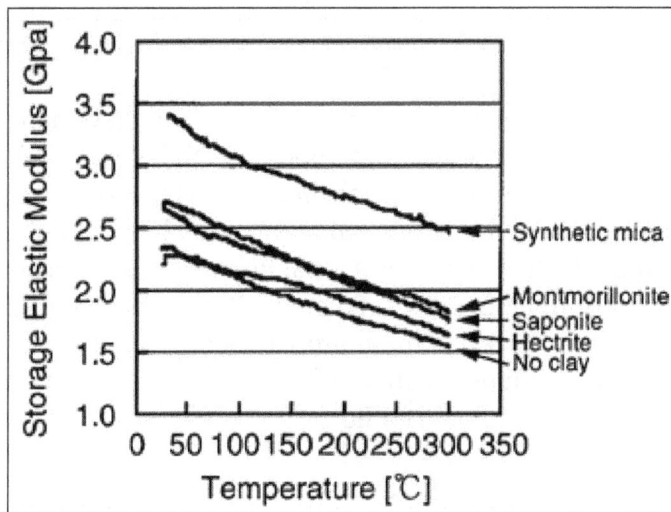

Variation of storage elastic modulus with temperature of polyimide-based nanocomposites filled with 2 wt% of nanoclay.

Polyimide/clay nanocomposites show enhanced stress and elongation at failure. Clay with montmorillonite (MMT) exchanged by hexadecylammonium with loading up to 5 wt% increases the bulk properties. Above this loading, the properties degrade by increasing brittleness. Increase in storage modulus is seen at any temperature as shown in figure. for polyimide-based nanocomposite filled with 2 wt% of organic clay. With such an improvement, glass transition temperature also decreases with increasing clay content. A greater improvement in mica-based nanocomposite is seen due to silicate layers with length of ~218 nm and thickness of ~1 nm. Mica and MMT clays have exfoliated structures, whereas aponite has a partially exfoliated–intercalated structure and hectorite-based nanocomposite has intercalated morphology. Polymer/LS possess gas barrier properties due to tortuous (maze) path created by LS as shown in figure. The tortuosity factor T_F is related to silicate length l_s, silicate thickness W_F and volume fraction of the nanocomposites V_F by,

$$T_F = 1 + \frac{l_s}{2W_F} V_F.$$

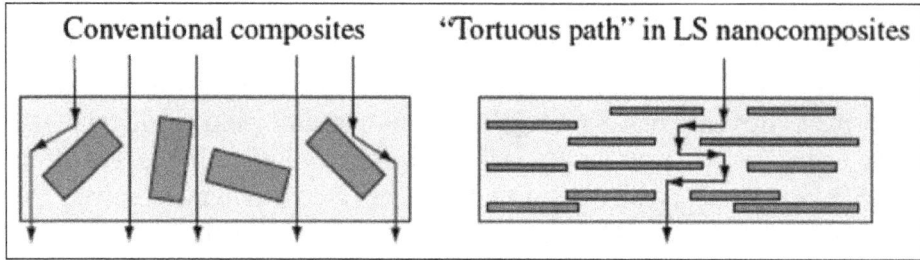

Tortuous path in polymer/LS nanocomposite.

The relative permeability coefficient P_R is the ratio of permeability coefficient of composite and polymer matrix, respectively. Relative permeability coefficient is found to be inverse of tortuosity factor $P_R = 1/T_F$. Thus, permeability coefficient is smaller if the length of silicate is longer. Tortuosity effects the permeability of nanocomposite PSPS by the relation,

$$P_S = P_P \left(\frac{1-V_F}{T_F} \right),$$

where P_P is the permeability of the pure polymer. Bourbigot et al. have demonstrated reduction in the permeability of gases like oxygen (O_2), water vapor (H_2O), helium (He), carbon dioxide (CO_2) and ethyl acetate vapor in clay-based nanocomposites. A 2 wt% clay loading decreases the water vapor permeability coefficient by 10-fold. Figure shows permeability of polyimide nanocomposites obtained with different types of fillers loaded by 2 wt%. Reduction in permeability is higher in nanocomposites with fillers of higher length. Fully exfoliated LS give the lowest permeability due to their creation of longer diffusive path. Owing to these properties along with excellent abrasion and heat resistance, polyimide/clay nanocomposites are used as step assist in automobiles. In a similar way, polyimide/clay nanocomposites are suitable for floor lining in the cargo area of commercial airliners.

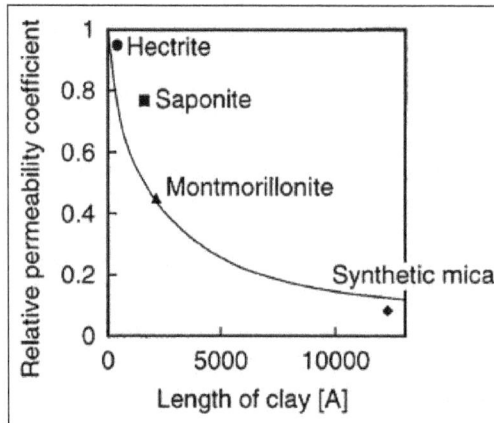

Variation of relative permeability coefficient with clay length.

Epoxy/Clay and Epoxy/CNT

Processing of polymer/clay or polymer/CNT nanocomposites involves the dispersion of clay or CNT in epoxy resin using ultrasonication followed by curing in molds to form a structure. Ultrasonication continues for usually 30 min followed by curing for 24 h. Curing is carried out at 70 °C to avoid damage to the nanostructures.

Mechanical testing done on epoxy nanocomposites with clay or CNT fillers reveals an increased modulus of elasticity as shown in table. Epoxy/CNT and epoxy/clay show better thermo-mechanical properties as compared to pristine matrix. COOH-functionalized epoxy/multi-walled carbon nanotube (epoxy/MWCNT) has a tensile strength of 121.8 MPa, whereas pure epoxy has a tensile strength of 95 MPa. Increase in CNT concentration increases the modulus of elasticity but decreases the tensile strength. Addition of nanoparticles is limited to 2% since rheological properties deteriorate above these levels.

Table: Mechanical properties of epoxy/clay and epoxy/CNT nanocomposites.

Nanocomposite	Tensile strength (MPa)	Shore hardness	Thermal stability (°C)	Modulus of elasticity (GPa)
Epoxy	95	75	50	2.8
Epoxy + 2 wt% 30 B	102.5	83	56	3.1
Epoxy + 2 wt% MWCNT	98	77	55	–
Epoxy + 2 wt% MWCNT-COOH	121.8	85	59	3.42
Epoxy	52.5	–	–	2.52
Epoxy + 0.1 wt% MWCNT	24.7	–	–	2.42
Epoxy + 0.5 wt% MWCNT	29.5	–	–	2.97

Epoxy/clay can serve as attractive replacements for the expensive talc powder and TiO_2 to attenuate the permeability in fuel tanks. Nanocomposites like fluoropoly(ether-amic acid)/montmorillonite (6F-PEAA/MMT) and polyethylene terephthalate/LS have also yielded significant improvement in mechanical properties, impact strength, barrier properties and flame properties. These properties are highly desirable in providing safety to aerospace structures in the event of fire and impact. Toyota used polymer/clay nanocomposite for structural applications by modifying natural clays. Nanostructured LS is a clay used to prevent gas diffusion through polymers. Thus, epoxy/clay nanocomposite is suitable for fabrication of extremely lightweight and strong cryogenic gas storage tanks for aerospace applications. These nanocomposites are also flame resistant which make them suitable for aerospace and space environments.

Nylon-6/layered Silicate (Nylon-6/LS)

Nylon-6/LS like nylon-6/MMT is prepared by a process of in situ intercalative polymerization. This involves the melt mixing of LS with nylon matrix without the need of a solvent. The LS is encased within liquid monomer to polymerize by the action of heat, radiation, diffusion, catalyst and exchange of cations between the layers. 2:1 phylosilicates is commonly used which has layer thickness of 1 nm made up of a sheet of alumina sandwiched between two silica sheets. Weak van der Waals interactions hold the layers together. Small molecules can be intercalated between these layers easily. Isomorphic substitution in the interface produces negative charges counterbalanced by the alkaline cations. Cations also provide functional groups that can react with polymer matrix increasing the strength of the formed interface. Montmorillonite $(M_x(Al_{4-x}Mg_x)Si_8O_{20}(OH)_4)$ (MMT), hectorite $(M_x(Mg_{6-x}Li_x)Si_8O_{20}(OH)_4)$ and saponite $(M_xMg_6(Si_{8-x}Al_x)O_{20}(OH)_4)$ are the examples of commonly used LS, where hydrated cations are ion-exchanged with bulkier organic cations. Montmorillonite imparts remarkable changes in thermal and mechanical properties of the nanocomposite. Incorporation of 4.1 wt% MMT in Nylon-6 matrix results in 102.7% increase in Young's modulus. Figure shows the variation of Young's modulus with clay content of an organo-modified nanocomposite. Nylon-6/LS are used in timing belt cover of automotive engines. These nanocomposites additionally possess an ablative property with which they have the potential for their use in aerospace engine components, body panels and crash guards.

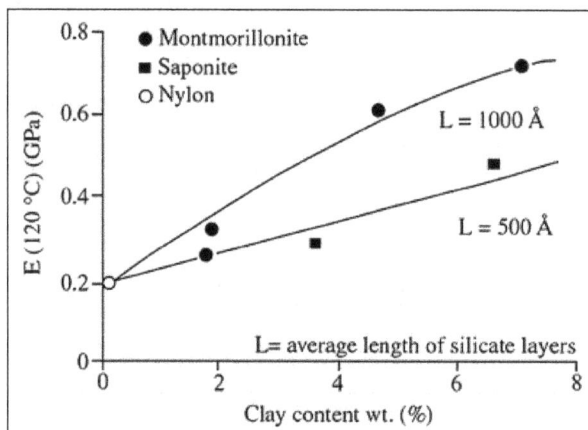

Variation of Young's modulus at 120 °C on clay content with organo-modified LS.

Epoxy/DOPO-based Phosphorus Tetraglycidyl

High-performance materials possessing superior thermal and mechanical properties have seen rapid developments for aerospace applications involving adverse operating conditions. Epoxy resins have excellent properties like toughness, adhesion and chemical resistance. However, epoxy does not possess thermal and mechanical properties required by the aerospace structures. Phosphorus containing flame-retardant compounds like triphenyl phosphate requires high loading to achieve the desired flame retardation. Polyhedral oligomeric silsesquioxanes (POSS) reagents have appeared as alternative materials, which can be incorporated easily in a polymer matrix like acrylics, styryls, epoxy and polyethylene.

A high-performance 10-dihydro-9-xa-10-phosphaphenanthrene-10-oxide (DOPO)-based phosphorus tetraglycidyl epoxy nanocomposite has been proposed by Meenakshi et al. for aerospace structural applications due to its good mechanical, thermal, flame-retardant and water absorption properties. The nanoreinforcement POSS amine was mixed with the epoxy resin followed by curing process. The mechanical properties of such cured nanocomposites with resin types containing nanoclay and POSS amine as nanofillers are compared in table Bis (3-aminophenyl) phenylphosphine oxide (BAPPO) and diaminodiphenylmethane (DDM) were the curing agents used. The mechanical properties of the nanocomposites with POSS showed better mechanical properties even when compared to the properties of clay-based nanocomposites. The reason was attributed to the small size of POSS that restricts the mobility of surrounding polymer chain and good interfacial adhesion. Flame-retardant tests showed enhancement in flame-retardant properties due to the protection of underlying matrix by char formation on surface and the low surface energy associated with the Si–O–Si present in POSS. Water absorption drastically reduces due to hydrophobic and partially ionic nature of Si–O–Si link. These exceptional mechanical, thermal, flame-retardant and water-barrier properties makes the polymer/POSS nanocomposite suitable for high-performance aerospace structures.

Table: Mechanical properties of epoxy/DOPO nanocomposites.

Resin system	Tensile strength (mpa)	Tensile modulus (gpa)	Flexural strength (mpa)	Flexural modulus (gpa)	Impact strength (KJ/M²)
TG-DOPO + DDM	91	6.90	148	4.78	35
TG-DOPO + 5% clay-DDM	77	7.00	157	4.88	31
TG-DOPO + 5% POSS-DDM	104	7.13	165	4.95	27
TG-DOPO + BAPPO	97	6.98	154	4.84	27
TG-DOPO + 5% clay-BAPPO	83	7.11	163	4.93	29
TG-DOPO + 5% POSS-BAPPO	110	7.20	171	5.05	24

Aerospace Coating Applications

Aerospace structures are subjected to diverse environments that include variations in moisture and temperature. They are also subjected to the contact with jet fuel, deicing fluid and hydraulic fluid. The coatings should withstand lightning strikes, ultraviolet exposure and erosion from dust flowing at 500 mph. An aerospace coating typically has a primer and a topcoat. The primer provides the substrate with adhesion and protection from corrosion. The topcoat is required to have matte finish, flexibility, durability, chemical resistance, corrosion protection and stable color for good appearance. These requirements are classified as shown in figure. The conventional ways to meet such requirements are explained in brief next.

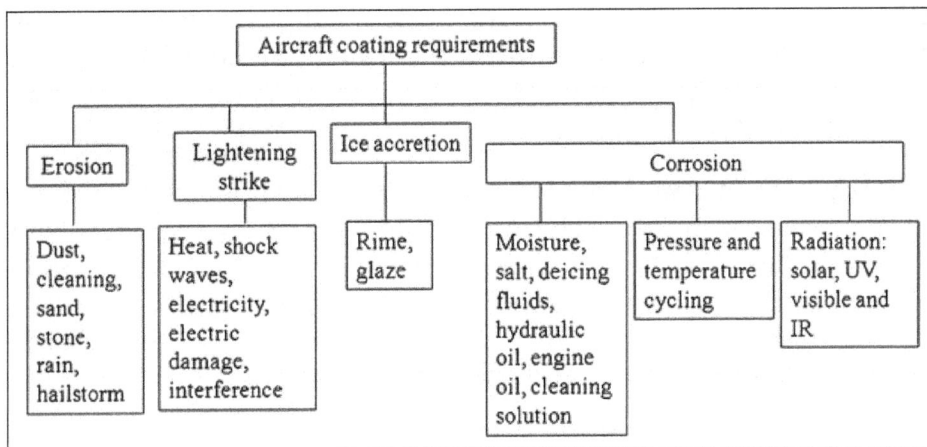

Requirements of aircraft structural coatings.

1. Lightening protection and electromagnetic shielding: Lightning strike can cause direct effects, indirect effects and damages in an aircraft. Direct effects and damages can be caused from heat (~30,000 °C), acoustic shock waves (500 psi impact force) and electrical charge (~200,000 A current). Indirect effect may cause damage or interruption of electronic equipment. Heat is generated by resistive or joule heating that causes most damage likely by the breakdown of matrix in composite. To prevent localized damage, thin metal mesh is embedded in outer layers of composite. This also serves as a shield for electrical systems. Nonconductive composites are given protection by strips on exterior surface to intercept lightning allowing the transparency to electromagnetic waves emitted from the antennas (through composites panels of antenna fairings). When radio frequency transparency is not required, conductive material is applied over exterior surface. To improve electromagnetic interference (EMI) and lightning strike resistance, conductive resins and

single-walled carbon nanotube Bucky-paper materials were introduced in composite materials. These Bucky-paper materials consisted of highly loaded and magnetically aligned CNTs whose conductivity changed linearly with temperature.

2. Erosion: Erosion due to sand particles is higher at oblique incidence due to scrubbing, micro-cutting action and chip formation in ductile metal and polymers and due to cracking and brittle fracture in ceramics and brittle polymers. Rain droplets cause impact induced debonding of coating from the surface causing fatigue and tearing. Rime and clear glaze of mixed ice can be rougher than a coarse sand paper during their peeling off from the aircraft skin causing erosion. Nickel, titanium shields, pure PU, blend of PU or polyethylene coatings are used conventionally for protection from erosion.

3. Deicing and anti-icing: Deicing and anti-icing performed on aircrafts help to remove and prevent the formation of ice. A thermally insulating coating reduces the efficiency of deicing process carried out by electrical heaters. In addition, retaining of heat within the insulation degrades the composite structures. Electric matrix, CNT paper-resist heaters and transparent MWNT films have been developed for deicing. Whereas, to reduce ice adhesion, ice-phobic coatings like Rain-X and MP55 poly(tetrafluoroethylene) (PTFE) coatings have been developed.

4. Radiation and corrosion protection: Ultraviolet rays, humidity, heat and gas present in atmosphere cause corrosion. Galvanic corrosion also results from composite (anodic) and metal combinations (cathodic). Since corrosion is a primary mode of failure in aircrafts, coating is designed to provide corrosion protection. Coatings with low permeability for liquids give barrier protection. Inhibitive pigments added in the coatings impart electrochemical property and passive protection to provide sacrificial protection.

To achieve these requirements, two-component PU systems have been used. Current aircraft coatings have limited life and capabilities. Polymer nanocomposites provide the flexibility to tune for mechanical, chemical, electrical, thermal, optical properties and their selective combinations for the aerospace coatings. Development of novel coatings to address the challenges of lightning, icing, erosion and corrosion is a focused area of research at present. Following subsections present the nanocomposite coatings and their specific aerospace applications.

Conductive Polymer/CNT for EMI Shielding

Nanocomposite coatings embedded with fillers of various functionalities suit for the topcoat of the aircrafts satisfying its multiple requirements. Polyurethane elastomers possess high wear resistance, corrosion resistance, elasticity and adhesive property. Fillers like clay, alumina, CNT and nanodiamond serve as main ingredients in the development of aerospace coatings. Spray painting is a technique adapted for aircraft painting. Adding fillers increases the viscosity causing difficulty in spray painting along with inadequate wetting of nanoparticles and additives. Alternatively, electrospinning method is used to fire a jet of solution through capillary tube on the surface. The possibility of adding nanofibers and CNT in the solution can produce coatings with good flexibility, strength and area-to-volume ratio. The fabricated coatings for aerospace applications differ in composition depending upon the structural location and its corresponding functional requirements.

Conductive filler like SWNT as low as 0.5 wt% is sufficient to impart electrical conductivity for electrostatic dissipation and EMI shielding. Variation of electrical resistivity with SWNT content (wt%) in a polycarbonate (PC) matrix is shown in figure. A drop of resistivity from 10^{13} to 10^8 is seen by an addition of 0.5 wt% MWNTs. For an addition of 5 wt% MWNT, the resistivity further reduces to 10^6. Resistivity at 0.5 wt% loading is sufficient to achieve the electrostatic discharge requirements. Such low loading of filler does not alter other physical properties of the coating. Aircrafts that are heavily equipped with electronic systems can seek protection from high-intensity radiated fields by sources in air or on land or sea by employing PC/SWNT nanocomposite coatings. In addition, the surface conductivity of PC/SWNT nanocomposite coatings also enhances the lightning strike protection.

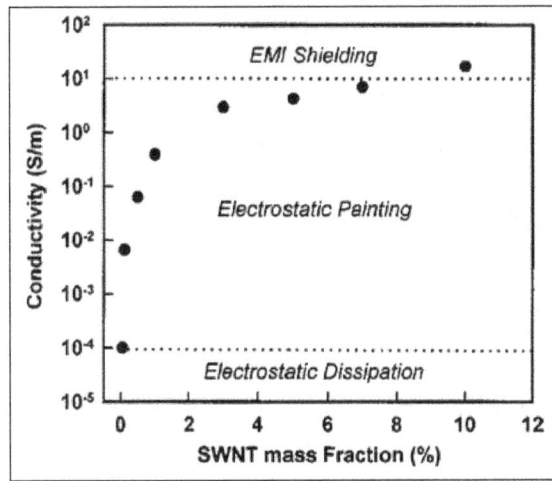

Variation of electrical conductivity with SWNT loading of a PC/SWNT nanocomposite (dashed lines represent the lower limit/percolation threshold).

Conductive Epoxy/CNT Heaters for Deicing

Image from scanning electron microscopy (SEM) showing the morphology of densified CNT film and an optical image of a small film on a Teflon sheet.

Ice accretion occurs on an aircraft flying between 3000 and 5000 ft, consequently increasing the weight and drag. Ice accumulation significantly alters the stall speed, which is risky. To mitigate ice accumulation, ice protection systems (IPS) in aircrafts use the air bleed from engines, which are not economical. Electromechanical deicing using ultrasonic exciters and electromagnets causes aircraft skin fatigue and noise. Electric heaters are widely used to prevent icing with greater efficiency. These solutions available today have limitations of weight, power, maintenance and integration within the structure. Thin-film nanocomposite heating element (0.038 mm) has been developed which is strong and flexible. These heaters have negligible weight, consume low power, are maintenance free and can be easily integrated within the structure. The heater patch consisted of ~100 μm tall CNT arrays synthesized by a thermal catalytic chemical vapor deposition process. The CNT array was densified to a film by pushing them in one direction using a steel tool of small radius. The film and its morphology are shown in figure. The films were integrated on a bistable composite laminate and on an aero-surface substrate using for testing. The dominant charge transport mechanism in such films is tunneling, which causes heating.

The voltage was applied across the heater on a bistable laminate, and the current was measured across it. A linear relation between voltage and current was seen above 0.1 A, where the resistance stayed constant. At 0.63 A and 5 V, the bistable laminate snapped due to the rise in temperature. However, due to heating, the resistance did not change indicating the thermal stability of the nanocomposite heater. The snapping of the laminate causes the curvature to change rapidly causing the ice to shear off from the surface especially during a cold start. The heating and snapping of the wing surface ensure the freedom from ice buildup. The power P_H required by thin-film heater drawing current I_H is inversely proportional to thickness as,

$$P_H = \frac{I_H^2 R_H L_H}{w_H h_H},$$

where R_H, L_H, w_H, and h_H are the resistivity, length, width and thickness of the element. High electrical conductivity and low thickness allow the heater to heat itself and the substrate quickly. A linear relationship between power and temperature (of the heater and laminate) was seen above 0.1 W of power. Thus, deicing temperature can be controlled by controlling the power of the nanocomposite heater. Ice tunnel test was performed on wings with nanocomposite heaters embedded on the skin. Harsh environment was created with temperature from 20.9 to −3.9 °C, with an airspeed of ~55.9 m/s and water content of 1.1 g/m³ with mean volume droplet size of 30 μm. Nanocomposite heaters demonstrated successful deicing capability as required by aircrafts flying through snow and storm.

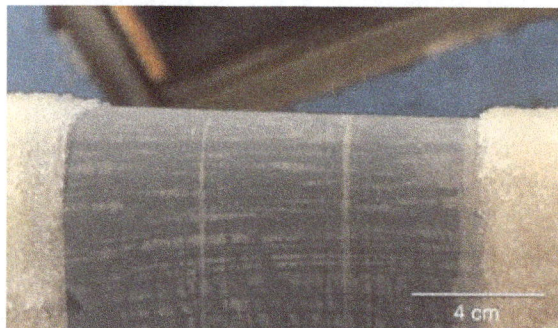

Aerosurface showing deicing during an IPS testing in ice tunnel.

Scratch-resistant Epoxy/Clay and Epoxy/Alumina

Epoxy coatings are low-cost coatings exhibiting good hardness, adhesion, chemical resistance and corrosion resistance. Turri et al. developed nanostructured coatings made of epoxy/clay and epoxy/alumina having additional properties like greater scratch strength, thermal properties and barrier properties, fulfilling major requirements of aerospace coatings. Fabrication of these coatings involves the mixing of resin and nanoparticles at 40 °C for 30 min. Nano-Al_2O_3 (AluC) and cloisite 30B, organophilic MMT clay (CL30B) fillers were added to prepare two different types of nanocomposites. Hardener was added to the mixture, and the resulting mixture was bar coated on a glass substrate followed by curing consecutively at ambient temperature, at 50 °C and at 120 °C, for 2 h at each temperature level to get a coating of 200–400 µm. Self-cross-linked hybrid coatings were prepared by mixing 5% silica in a silanized epoxy. Silanized epoxy was prepared by blending diglycidyl ether of bisphenol A and 3-aminopropyltriethoxysilane in 1:2 molar ratios for 2 h in nitrogen atmosphere. This mixture was added in a sol containing tetraethyl orthosilicate, water and ethanol at a molar ratio of 1:3:4.5 and gently stirred. Similarly, epoxy/Al_2O_3 coatings were also prepared.

Atomic force microscopy (AFM) topography and demodulation images were obtained to study the surface roughness and scratch hardness. The surface topography was obtained from the mean surface roughness and root-mean-square values. A demodulation phase image for the hybrid epoxy/SiO_2 nanocomposite is shown in figure. The roughness was estimated from median surface level and standard deviation within the image. The small domains indicated by bright spots are the hard phase, and dark zones are the soft polymer matrix. Phase separation is not seen in the image, and silica particles were fully embedded in the polymer matrix.

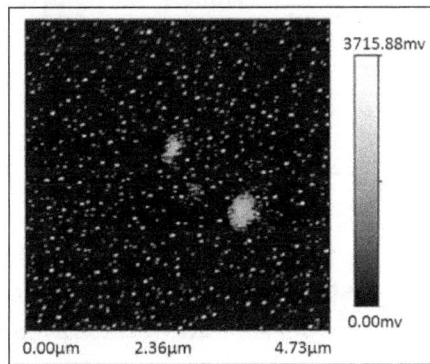

Phase image of epoxy/SiO_2 nanocomposite coating.

Topographic images obtained after making a nanometric scratch on epoxy/clay and on hybrid epoxy/SiO_2 coating are shown in figure. A constant force of F_{AFM} = 3400 nN was applied on the AFM tip. Pileup was observed at the scratch location from which the width of the scratch was measured. Scratch hardness H_S is given in terms of normal load F_{AFM} applied on the tip of AFM and width of the scratch W_S as,

$$H_S = \frac{4F_{AFM}}{\pi W_S^2}.$$

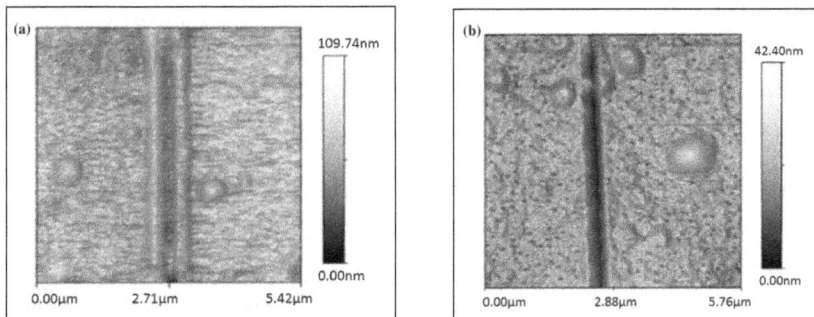

Topographic images of: a epoxy/clay nanocomposite coating and b hybrid epoxy/SiO_2 nanocomposite coating.

Standard tests like the Buchholtz indentation (for indentation hardness) and Taber tests (for wear resistance) were conducted. The indentation hardness (evaluated from Equation $H_S = \dfrac{4F_{\text{AFM}}}{\pi W_S^2}$.) and wear resistance estimated for various types of coatings are given in table Alumina and clay-based nanocomposites showed improvement in indentation and scratch hardness when compared to a pristine epoxy. However, there was considerable increase in surface mean roughness of epoxy/clay nanocomposite. The wear index was improved in case of epoxy/alumina and decreased in case of epoxy/clay. Thus, epoxy/alumina coating is better than the epoxy/clay coating. The hybrid epoxy/SiO_2 coatings showed a remarkable improvement in scratch hardness maintaining the surface roughness below that of epoxy/clay coatings. Very high improvement in scratch hardness of 1830, 942 and 700% was seen in hybrid epoxy/SiO_2 coatings when compared to pristine epoxy, epoxy/alumina and epoxy/clay coatings, respectively. Thus, hybrid epoxy/SiO_2 coatings can be proposed as coatings for aircrafts operating in dusty conditions, sandstorms and hailstorms.

Table: Properties of various Epoxy Nanocomposite Coatings.

Coating type	Mean surface roughness (nm)	Indentation hardness	Wear index (mg/kcycle)	Scratch hardness (MPa)
Epoxy	0.16 ± 0.02	99 ± 2	63 ± 15	42.8 ± 3.2
Epoxy/alumina	0.59 ± 0.04	101 ± 7	75 ± 13	83.1 ± 3.1
Epoxy/clay	3.15 ± 0.04	98 ± 3	33 ± 5	112.3 ± 5.6
Hybrid Epoxy/SiO_2	2.65 ± 0.09	87 ± 11	35 ± 7	783.6 ± 51.6

Hydrophobic PVDF/MWCNT

Superhydrophobic coatings have anti-sticking, anti-icing, self-cleaning and anti-corrosion properties, which can provide fluidic drag reduction and water corrosion prevention to the aerospace structures. Key parameters governing hydrophobicity are the nanoroughness and surface energy. Wenzel model describes an increased surface roughness leading to increased surface area imparting hydrophobic nature, whereas the Cassie–Baxter model states that the air trapped within the grooves causes hydrophobic behavior. Poly(vinilidene fluoride) PVDF is a strong polymer having properties like abrasion resistance and corrosion resistance.

Addition of CNTs imparts additional properties to the PVDF like hydrophobicity and conductivity (for EMI shielding and lightning protection) making the PVDF/CNT nanocomposites a suitable material for aerospace coatings. PVDF/CNT nanocomposites coatings were prepared by Chakradhar et al. using spray coating method. PVDF and MWCNTs were dispersed in dimethyl formamide (DMF) using a magnetic stirrer for 10 min, followed by ultrasonication for 50 min. The mixture was mixed with Milli Q water resulting in a precipitate. Precipitate was collected on a filter paper and dried. The dried powder was dissolved with acetone by ultrasonication for 30 min. The mixture was sprayed on a glass substrate to form 10–12 μm thick coatings.

Surface morphology of coating with 66 wt% MWCNT examined by field electron scanning electron microscopy (FESEM) is shown in figure. Protrusion-like structures were seen which cause enhanced surface roughness leading to reduced wettability as per the Cassie–Baxter's law. The PVDF/MWCNT coatings exhibit superphobicity in contact with the water. The water contact angle (WCA) and sliding angle (SA) were found to be 154° and <3°, respectively (superhydrophobic), as compared to the WCA of PVDF which is 105° (hydrophobic). The nanoroughness of the coating increased with the addition of MWCNT causing superhydrophobicity. WCA measured at different temperatures is shown in figure. For 20–30 wt% coatings, the WCA gradually reduced reaching 10° at 623 K. The 33–66 wt% coatings showed considerable stability up to 573 K. Thus, 33 wt% of MWCNT in the PVDF imparted the necessary superhydrophobicity to the coatings maintaining additional properties of the PVDF. These properties make the PVDF/MWCNT coatings suitable for aerospace coatings for wet climates.

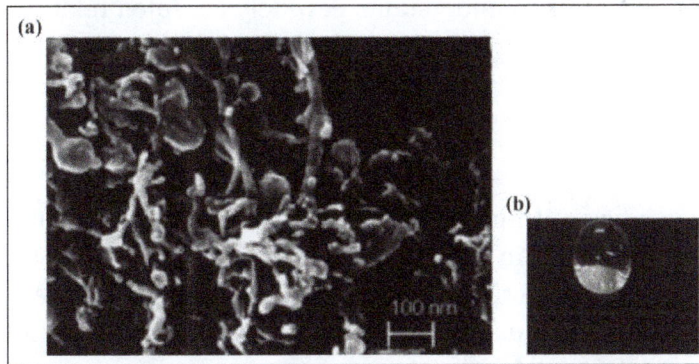

(a) Field electron scanning electron microscopy (FESEM) image of a PVDF/MWCNT coating with 66 wt% MWCNT at room temperature. (b) Water droplets on a PVDF/MWCNT superhydrophobic coating at 300 K.

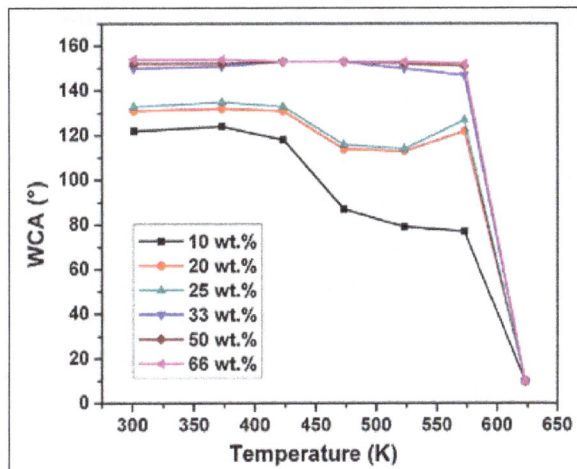

Variation of WCA with temperatures in PVDF/MWCNT coatings.

Aerospace Tribology

Addition of lubricating and reinforcing fillers imparts tribological properties to the nanocomposites. Lubricating microfillers like graphite, PTFE, molybdenum disulfide (MoS_2) decrease the friction of the sliding surfaces, but their weaker bond in the material reduces the strength. Reinforcing microfillers like glass and carbon fibers increases the strength but also increases the abrasiveness. Nanofillers overcome the drawbacks of these microreinforcements by the improvements in some characteristics of the nanocomposites. Nanofillers increase the nucleation capability by improving interfacial interaction with the polymer matrix. Rigidity of the matrix is increased due to restricted mobility of the matrix occurring near vicinity of the fillers. Nanofillers are more effective than microfillers due to the increased surface area of fillers that create huge polymer/filler interfacial area resulting in enhanced bonding. In addition, material removal during sliding is less since the size of nanofiller and surrounding polymer chain is similar.

Apart from changing the physical and mechanical properties, nanofillers change the crystallinity, microstructure, glass transition temperature and degradation temperature. Nanofillers influence the transfer films, which is crucial in minimizing the wear rate in bearings. Wear results from the rubbing bodies due to friction, impact, thermal and other forces. Debris dislodge from the surface due to the loss of cohesion causing wear mechanisms like abrasion, adhesion, erosion, delamination, corrosion and fatigue. These mechanism involves softening, melting and pyrolysis making wear phenomenon more complex to understand. A widely accepted relation to estimate the wear volume loss V (m³) is given by,

$$V = \frac{k_{\mathrm{w}} \mathrm{NL}}{H},$$

where k_{w} is the wear coefficient (mm³/N-m), N normal load (N); l sliding distance (m); and H hardness of the wearing material. An important factor governing wear is the tribochemical reaction. Thus, Gibbs free energy (ΔG) criterion facilitates the selection of fillers. A negative ΔG indicates feasibility of a chemical reaction that helps on the formation of tribofilm reducing wear and favoring them for tribological applications.

Shao et al. found that the frictional coefficient and wear rate decreased rapidly when 1 vol% nano-TiO$_2$ particles were added to PPESK copolymer. Nanocomposite was slided over a steel ring in dry condition with velocity = 0.43 m/s, load = 200 N and duration = 1.5 h. Micro-TiO$_2$ particles had negligible effect on friction coefficient and resulted in the increasing wear rate with increasing filler content. A continuous thin transfer film well bonded to the surface decreased the wear rate. A thin Fe$_2$O$_3$ film was formed, and the elemental Ti got attracted toward the steel surface bonding the formed film to the surface. PTFE nanocomposites showed the smooth and coherent transfer films. Increasing the filler content increased the abrasiveness thereby increasing the wear rate. Si$_3$N$_4$ also exhibited hydrodynamic lubrication behavior due to tribochemical wear when used in thermoplastic or thermosetting polymer. A formation of SiO$_2$ in the tribofilm reduced the friction and protected the sliding surfaces. Table lists the tribological properties of some nanocomposites. The lowest wear rate was exhibited by polyimide/CNT nanocomposite. These PMNCs are lightweight, have low friction and have low wear rate. In addition, fillers like CNT impart thermal conductivity to remove the heat built up during operation. Thus, PMNCs are the preferred materials for the aerospace bearing design.

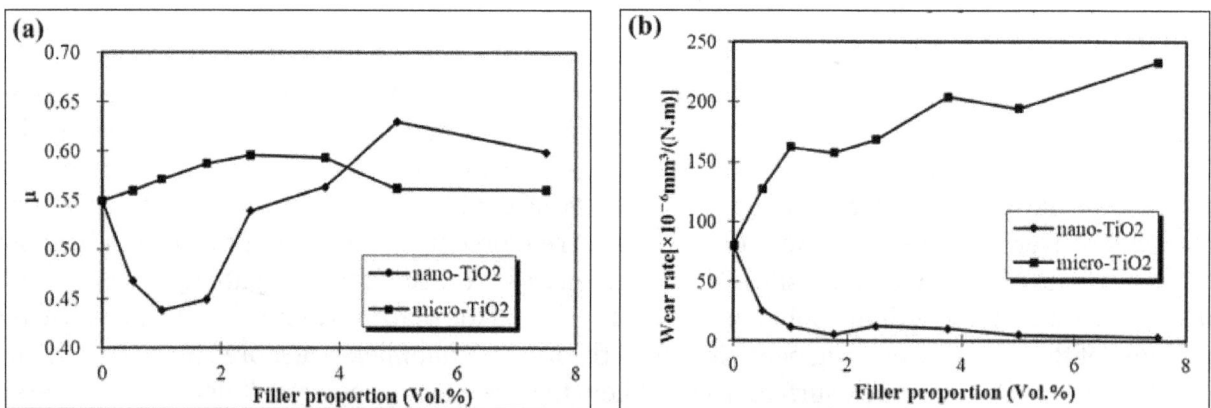

Variation in properties of nano-and micro-TiO$_2$-reinforced PPESK with different filler volume proportions: a coefficient of friction and b wear rate.

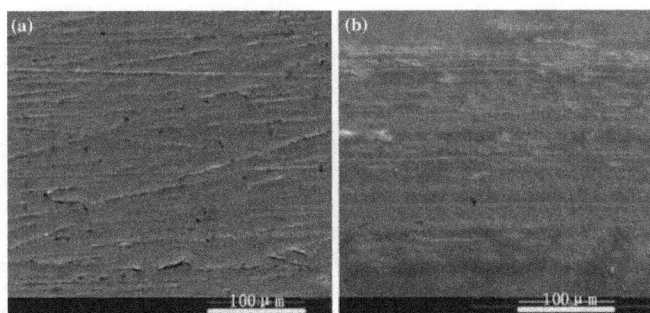

Micrographs showing scanning electron microscopy (SEM) images of transfer films, a pure PTFE and b 17 wt% TiO$_2$ particles in PTFE. Results obtained were with load of 200 N, sliding velocity of 1.39 m/s, sliding time = 30 min, sliding distance = 2.5 km.

Table: Tribological properties of some polymer nanocomposites.

Material systems	Wear rate (10^{-6} mm³/km)	Toughness (kJ/m²)	Hardness (MPa)
Epoxy	38	9.3	19
Epoxy/4 vol% TiO$_2$	15	46	–
Epoxy/0.9 vol% nano-Si$_3$N$_4$	7	13.2	20
Epoxy/2 vol% Al$_2$O$_3$	3.9	55	–
Epoxy/3 vol % SiO$_2$	22	–	208
Polyimide/10 wt% CNT	0.1	–	380
Polyimide/10 wt% graphite	0.7	–	300
PEEK/7.5 wt% Si$_3$N$_4$	1.3	–	133
PEEK/7.5 wt% SiO$_2$	1.4	–	–
PEEK/2.5 wt% SiC	3.4	–	–
PEEK/10 wt% ZrO$_2$	3.9	–	–
PPS/5 vol% TiO$_2$	8	–	–
PTFE/A$_{12}$O$_3$	1.2	–	–

Structural Health Monitoring Applications

Structural health monitoring (SHM) systems increase the reliability and safety of structures, which is mandatory for many critical and high-performance structures of aircrafts. SHM systems have reduced the cost involved by periodic inspection, which happens to be around one-third of the total cost in acquiring and operating the aircrafts. Piezoelectric and piezoresistive nanocomposite sensors and actuators have been developed recently that are easy to fabricate, have low cost, making them suitable for the use in SHM systems of large aircrafts and various measuring gauges.

Piezoelectric Polymer Matrix Nanocomposites

Piezoelectric PZT wafers are the widely used transducer elements in SHM. Due to their toxic nature, brittleness and high cost, piezopolymer transducers like PVDF and poly(vinylidene fluoride–trifluoroethylene) (P(VDF-TrFE)) have been developed as alternatives. To facilitate the processing, piezoelectric PMNCs have been developed. Nanocomposites also possess structural and electrical properties that can be tailored by controlling its composition to make them suitable for transducer applications. Piezoelectric PMNC consists of piezoelectric nanofillers like PZT, zinc oxide (ZnO) embedded in a polymer matrix. These nanofillers impart piezoelectricity simultaneously retaining the properties of the polymer. Recently, PVDF and epoxy polymers have received close attention as matrix material.

PVDF/PZT Films

PVDF/PZT films exhibit piezoelectricity due to both PZT nanoparticles and the PVDF matrix. It also possesses the properties of PVDF like low acoustic impedance, strength, corrosion resistance, flexibility, nontoxic nature satisfying most of the requirements for aerospace sensor applications. Jain et al. have developed piezoelectric PVDF-PZT nanocomposite film sensors using a solvent cast method. A measured quantity of PVDF was dissolved in DMF by stirring and heating in the microwave oven until it dissolves. The PZT was dispersed thoroughly in the polymer PVDF solution. The average grain size of the PZT particles was in the range of 1 μm to 25 μm. The prepared solution was then casted on the glass plate and heated in an oven at about 65 °C. With the complete evaporation of solvent, PVDF-PZT nanocomposite films were obtained. The process was repeated with different weight fractions (ranging from 10 to 50 wt%) and different particle sizes of PZT. The prepared films were tested for structural, surface and mechanical properties by X-ray diffraction (XRD), SEM and tensile testing techniques, respectively. The graph in figure. presents the X-ray pattern of the film. It is clear from the patterns that most intense reflection occurs at $2\theta-30.98°$, which is very close to the value mentioned in the literature for PVDF-PZT nanocomposite. The intensity of reflection increases as the PZT concentration increases from 10 to 50 wt%. In addition, with the increase in size of PZT particle, the pattern remains same. This indicates that although the particle size of PZT is varied, the crystallite size remains same.

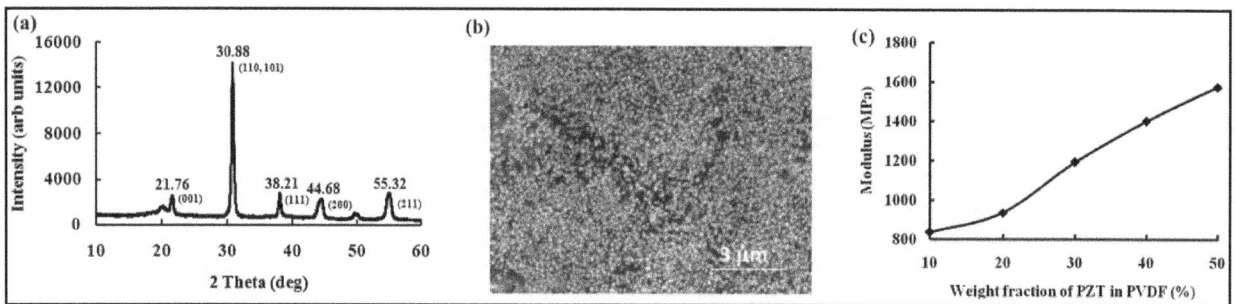

Material characterization of PVDF-PZT nanocomposite film: a XRD pattern, b SEM, c variation in modulus with weight fraction of PZT.

In order to study the microstructure and the dispersion of the ceramic powder within the polymer matrix, SEM was performed. A uniform distribution of PZT particles in the PVDF matrix is seen from figure. It has been found that the modulus increases with an increase in PZT content. At 50 wt% of PZT, modulus is found to be 1574 MPa. The increase in modulus and degree of crystallinity improves the piezoelectric properties. After synthesis, the PVDF-PZT composite films were electroded using silver ink and were poled using thermal and corona poling method by the application of high direct current (DC) electric field (9 kV/mm) at an elevated temperature. Piezoelectric coefficients have been measured using a Piezometer PM300. The average value of piezoelectric charge coefficient d_{33} obtained is 40 pC/N, and the maximum value obtained is 100 pC/N, which is much higher than the value obtained for the single-phase PVDF, which has an average d_{33} of 22 pC/N.

The PVDF-PZT sensor is tested for static and dynamic strain sensing properties. The voltage response of the film bonded on a beam is measured by subjecting the beam to free vibration and transient impact loading. The sensor bonded on a cantilever beam at its root is connected to the oscilloscope. First, the sensitivity of the bonded PVDF-PZT sensor to the dynamic strains induced from the free vibration of the beam is studied.

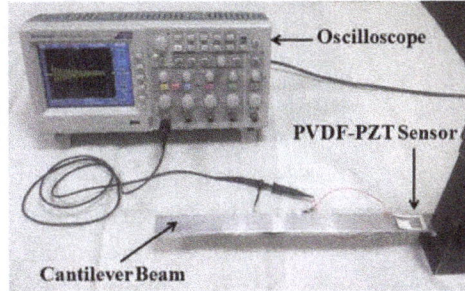

Experimental setup showing the cantilever beam with PVDF-PZT nanocomposite sensor bonded on it.

The cantilever beam specimen used is aluminum having length L = 0.33 m, width = 0.03 m and thickness h = 0.0015 m. A free vibration is induced in the beam by giving initial displacement and leaving it to vibrate on its own. The response of the PVDF-PZT sensor is recorded using oscilloscope with a sampling rate of 1 kHz. The voltage response for free vibration is shown in figure. The frequency response is obtained by taking fast Fourier transform (FFT) of the voltage response as shown in figure. Number of sampling points used in FFT is 2×10^{12}. The peaks reveal the resonance frequencies of the vibrating modes of cantilever beam. The first and second natural frequencies (F_1 and F_2) of the beam can be obtained using elementary beam theory (Euler–Bernoulli). The first two natural frequencies are given by,

$$F_1 = \frac{1.875^2}{2\pi}\sqrt{\frac{EI}{\rho AL^4}},$$

$$F_1 = \frac{4.694^2}{2\pi}\sqrt{\frac{EI}{\rho AL^4}},$$

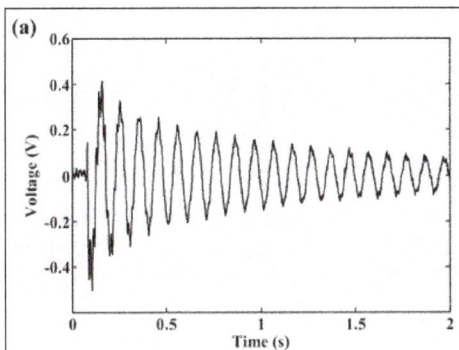

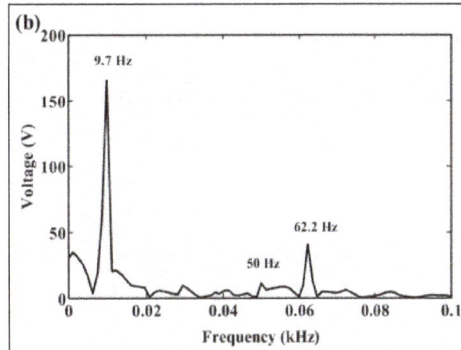

Response of PVDF-PZT bonded on a cantilever beam subjected to free vibration:
a voltage response, b frequency response.

where E is Young's modulus, $I = wh^3/12$ is the area moment of inertia, ρ is the density and A is the cross-sectional area. For aluminum material with $E = 60 \times 10^9$ and $\rho = 2700$, the first two natural frequencies estimated from elementary beam theory are $F_1 = 10.48$ and $F_2 = 65.73$ Hz. The natural frequencies from the graph are $F_1 = 9.7$ and $F_2 = 62.2$ Hz. The peaks are in good agreement with respect to the natural frequencies obtained from the elementary beam model. Thus, the developed PVDF-PZT nanocomposite film sensors have the ability to capture multiple modes of vibrations accurately.

Response of the PVDF-PZT sensor to impact loading is obtained by dropping a steel ball on the tip of the cantilever beam from a known height. A ball diameter of $d = 0.0095$ m is dropped from a height of $h_1 = 0.4$ m at the tip of the cantilever beam. The steel ball hits the beam tip and falls off with negligible rebound of 1–2 mm converting entire impact energy into mechanical vibration. Magnitude of impact in terms of energy can be estimated as $E_1 = mgh = 0.0137$ N-m. The mass of ball is found to be $m = 3.5$ g. Impact response of the sensor mounted at the root of the cantilever is shown in figure. The frequency response (FFT of the sensor signals) is shown in figure. The sensor effectively captures the peaks in the frequency response representing several resonant frequencies of the beam. The natural frequencies from the graph are $F_1 = 9.7$ Hz and $F_2 = 62.9$ Hz, which again closely matche the theoretical values. Thus, the developed PVDF-PZT nanocomposite films have the ability to sense dynamic strains resulting from impact. With good sensitivity, low cost and corrosion resistance, these films are suitable for acoustic emission monitoring and condition monitoring of aircraft structures.

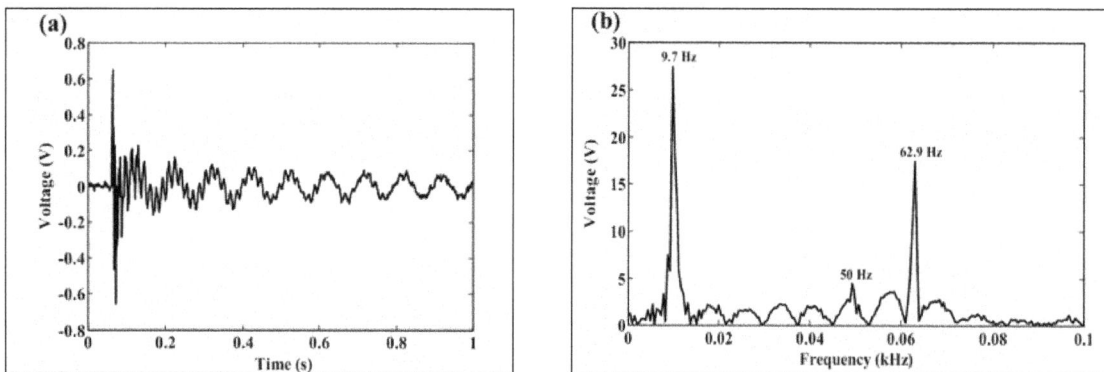

Response of PVDF-PZT nanocomposite sensor bonded on a cantilever beam subjected to impact of magnitude $E = 0.0137$ N-m at the cantilever tip. a Voltage response. b Frequency response.

Epoxy/PZT Coatings

The high PZT content in an epoxy matrix often leads to poor material homogeneity, which in turn lowers the mechanical properties. Recently, a potentially attractive class of epoxy called liquid crystalline thermosetting resins has been developed for the addition of PZT. The piezoelectric charge constants (d_{33}) as a function of the PZT loading fraction are depicted in figure. The piezoelectric constant increased with the increase in PZT loading. The novel epoxy/PZT nanocomposites produced possess a high piezoelectric charge constant d_{33}. Recently, Rathod et al. have used embedded epoxy/PZT sensors with 80 wt% PZT for acoustic emission sensing applications for composite laminates. The acoustic emission sensed due to fiber breakage during the tensile testing is shown in figure. Thus, the epoxy/PZT sensors qualify as the material for impact, vibration and acoustic emission sensing in aircraft structures. These nanocomposites show the potential to replace existing piezoelectric materials for automobile and aircraft applications especially operating in harsh environments.

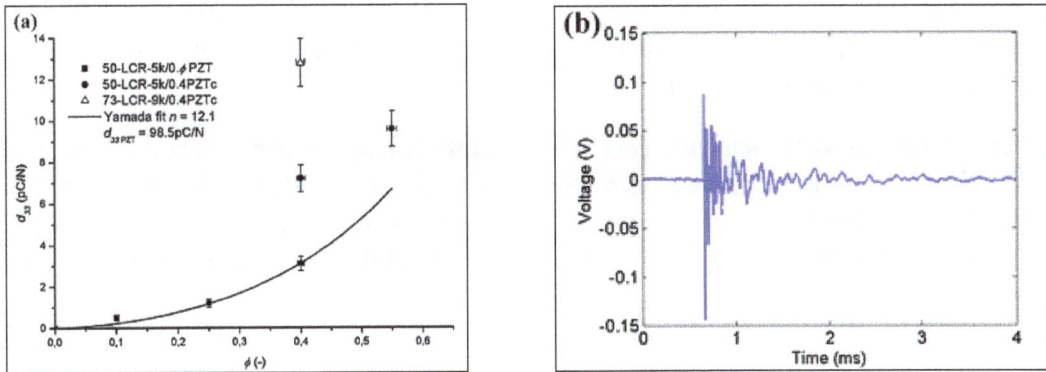

(a) Piezoelectric coefficient d_{33} of the nanocomposite material as a function of the PZT volume fraction ϕ. Yamada's model is fitted to the data. (b) First acoustic emission observed from embedded sensors under uniaxial stretching at 1 mm/min from embedded piezoceramic coating film.

Polymer/Zno Films

Apart from PZT nanoparticles, ZnO particles also exhibit inherent piezoelectricity developed ZnO-embedded piezopolymer thin films for dynamic strain sensing. Piezopolymer films were prepared by dispersing 1 wt% ZnO nanoparticles (20 nm) in PSS solution via 180 min of bath sonication (135 W, 42 kHz) followed by 30 min of high-energy probe sonication (4.76 mm tip, 150 W, 22 kHz). The resulting mixture was mixed in an equal amount of PVA solution followed by evaporation of solvent at 25 °C for 72 h. The resulting PSS-PVA/ZnO nanocomposite was peeled off from the substrate whose ZnO weight fraction WFWF can be calculated from the relation,

$$W_{F} = \frac{W_{ZnO}}{W_{ZnO} + W_{PSS} + W_{PVA}} \times 100 \setminus \%,$$

where W_{ZnO}, W_{PSS} and W_{PVA} are the mass of ZnO, PSS and PVA, respectively. The SEM image of a nanocomposite with $W_F = 50$ is shown in figure. The random dispersion of ZnO nanoparticles indicated adequate dispersion in the polymeric matrix. The sizes of agglomerates varied from 20 to 38 nm, which are lesser than twice the size of nanoparticles. Thus, agglomeration was found to be negligible.

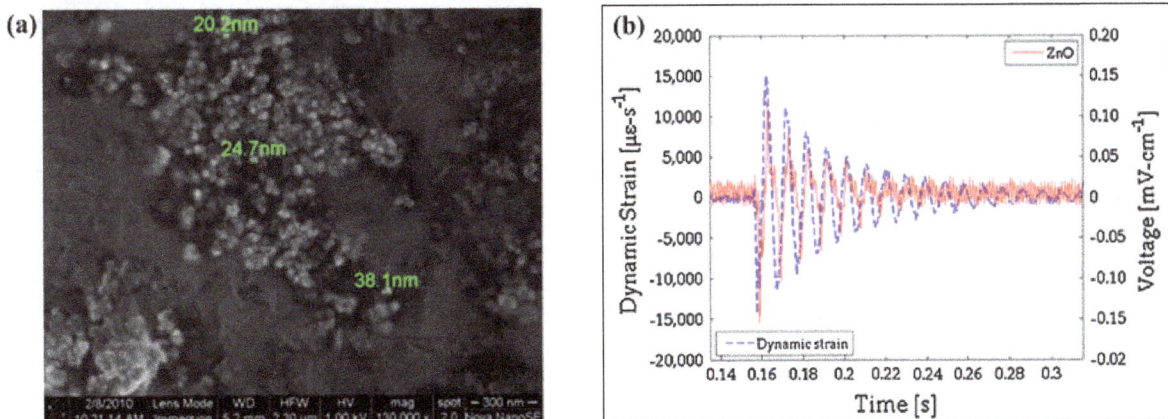

In figure presents a High-magnification SEM images of a PSS-PVA/ZnO thin film with 50% weight fraction of ZnO showing random deposition of individual and small agglomerations of ZnO nanoparticles (20 and 38 nm) within the polymeric matrix. b Free vibration response of a PSS-PVA/ZnO thin film with 50% weight fraction of ZnO overlaid with the induced dynamic strain as measured by the metal-foil strain gage.

The nanocomposite film was poled with electrodes mounted on opposite edges of the film under moderate electric field. The poled film was mounted on a cantilever beam near the fixed end. The cantilever was subjected to free vibration, and the voltage response of nanocomposite film was measured. The output voltage VOVO was normalized to take into account for the films of different sizes as,

$$V_F = \frac{V_O L_F}{Wh},$$

where L_F is the distance between the electrodes, W is the width of the film and h_f is the thickness of the film. The normalized voltage response of the film is shown in figure. The response is matched with the strain measured using standard strain gauge. These films are easy to prepare, and poling requires moderate electric filed. ZnO enhances the piezoelectricity in piezoelectric polymers. Realizing this, Meyers et al. have further demonstrated the use of ZnO-based nanocomposites as transducer for Lamb wave-based SHM in pipe structure. Thus, ZnO-based nanocomposites hold promising future in development of cost-effective SHM systems for aircraft structures.

Piezoresistive Polymer Matrix Nanocomposites

Piezoresistive sensors work on the principle of change in resistance caused by the application of external stimuli. Conventional strain gauges have a gauge factor of ~2, which is very low to detect the above-mentioned dynamic strain in aircrafts. These sensors and piezoelectric transducers prove costly to monitor large aircraft structures. The cost-effective piezoresistive nanocomposite sensors serve as an alternative to such conventional sensors. Piezoresistive nanocomposites contain conductive nanofillers like CNT and CB that imparts conductivity.

PMMA/CNT

Extremely small size of CNTs has enabled Kang et al. to develop artificial neuron made up of long CNT fiber-based PMNC. SWNTs were mixed in a solvent containing PMMA as a polymer-binding agent. Mixing was done using shear force at 70 °C for 4 h. Ultrasonic mixing was avoided since CNTs suffer damage by tip sonification and cavitation. The mix was cast in a Teflon mold and kept at room temperature to remove air, followed by curing in an oven at low vacuum (16 inch Hg), at 120 °C for 12 h. After evaporation of the solvent, a nanocomposite film of 85 μm was peeled from the mold. Sensitivity of the sensor was determined by bonding it near the cantilever root. For the controlled displacement y applied at cantilever tip, the strain ε is estimated by beam bending relation which is given by,

$$\varepsilon = \frac{3h(L-a)}{2L^3} y(L),$$

where h is the beam thickness, L is the beam length and a is the distance from beam tip to center of the sensor. Gauge factor G_f that estimates the sensitivity of piezoresistive sensor is given by,

$$G_f = \frac{\Delta R}{\varepsilon},$$

where ΔR is the normalized change of resistance $\Delta R = (R_s - R_I)/R_I$ with R_s being the final resistance after straining and R_I being the initial resistance. With a percolation threshold of 0.1%, the variation of gauge factor is shown in figure. where the gauge factor decreases as the CNT loading increases. The optimized content of SWNT was found to be 3 wt%. The MWCNT-based nanocomposite neuron possesses SHM capability which can be seen from the response observed when it is cut half way along its width. Thus, nanocomposite neurons can monitor very large areas of an aircraft for damages. Its easy fabrication, simple instrumentation and low cost make it an excellent choice as a sensor for SHM systems for large aircrafts.

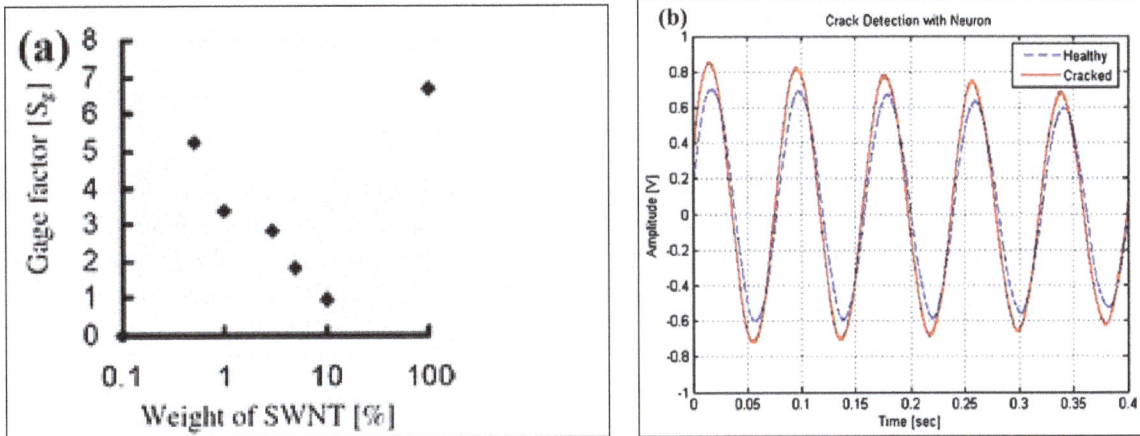

A Gauge factor of PMMA/SWNT nanocomposite strain sensor, b dynamic response of a MWNT neuron on a beam before and after crack, demonstrating the dynamic strain sensing capability.

Epoxy/Carbon Black and Epoxy/CNT

Embedded conventional sensors or transducers on the structure can act as a defect and create interference in the normal operation of the structure. For example, transducers (like PZT wafers) mounted on the surface of the skin of a wing may disturb the aerodynamic flow and reduce the efficiency of the wing. To address this problem, smart paints have been developed which are capable of sensing the dynamic strain apart from providing the protective and decorative functions of the paint. Smart composite paints developed by piezoelectric powder and epoxy resin require electroding and poling, which is complex and expensive. Realizing the ability to measure the static and quasi-static pressure, Aldraihem et al. demonstrated the ability to measure dynamic pressure using epoxy/carbon-black nanocomposite sensor.

The nanocomposite paint sensors were prepared by mixing CB particles (42 nm) in PU and pouring the mixture in metal mold for one day to form 0.942 mm thick films. Disk-shaped samples (24.6 mm diameter) were cut and electroded with silver paint as shown in figure. The prepared paint sensors were tested by an arrangement as shown schematically in figure. The force $F(t)$ was applied on the sensor using a shaker. The voltage response was obtained from a Whetstone's bridge. A single degree of freedom system model yields a set of coupled governing differential equations obtained by using the Hamilton's principle for electromechanical systems as,

$$M\ddot{x} + b\dot{x} + Kx - \frac{Q_{2dc}}{(\varepsilon A)}Q_2 = F,$$

$$(R_0 + R_a)\dot{Q}_1 - R_g(\dot{Q}_2 - \dot{Q}_1) = 0,$$

$$R_g(\dot{Q}_2 - \dot{Q}_1) + \frac{x_{dc}}{\varepsilon A}Q_2 - \frac{Q_{2dc}}{\varepsilon A}x = 0,$$

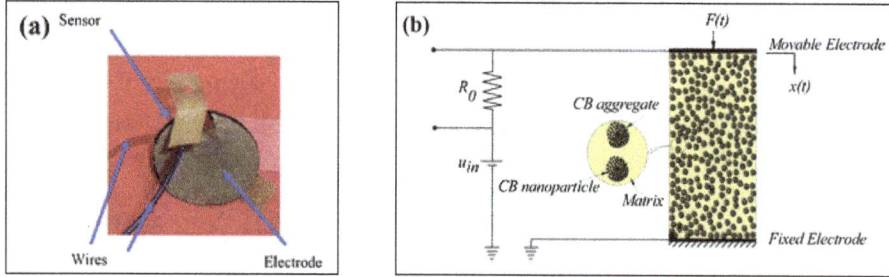

(a) Circular sample of PU/CB nanocomposite paint sensor. (b) Schematic arrangement of the paint sensor and its equivalent circuit.

where, M, K and b are the equivalent mass, stiffness and damping, respectively. x is the displacement of the equivalent mass. R_a, R_g, R_0 and C_g are the series resistance, parallel resistance, output resistance and parallel capacitance, respectively. Q_1 and Q_2 are the electrical charges created by excitation, and $Q_{2dc}/\varepsilon A$ is the electromechanical coupling factor. ε is the permittivity, and A_E is the electrode area. For a given $F(t)$,

$$M\ddot{x} + b\dot{x} + Kx - \frac{Q_{2dc}}{(\varepsilon A)}Q_2 = F,$$

Equation $(R_0 + R_a)\dot{Q}_1 - R_g(\dot{Q}_2 - \dot{Q}_1) = 0,$

$$R_g(\dot{Q}_2 - \dot{Q}_1) + \frac{x_{dc}}{\varepsilon A}Q_2 - \frac{Q_{2dc}}{\varepsilon A}x = 0,$$

can be solved for $x(t)$, $Q_1(t)$ and $Q_2(t)$. Voltage change $u\bar{\ }u\bar{\ }$ due to mechanical vibration is given by

$$\bar{u} = u_{dc} + u(t) = R_0\dot{Q}_{1dc} + R_0\dot{Q}_1,$$

where the first part is due to static pressure and the second part is due to mechanical vibration. After modeling of the nanocomposite paints, Aldraihem et al. superposed its results that matched well with the experiments as shown in figure. With the availability of mathematical model and nanocomposite paint sensor, the sensor sensitivity can be studied for tuning capabilities to suit for parameters like frequency, strain level and environment conditions encountered by aerospace structures. The epoxy/CB nanocomposite paint can act as a continuously distributed sensor over large aircraft structures, fairing of launch vehicles and flexible space structures for vibration, noise and health monitoring applications. Apart from these carbon black-based pressure sensors, strain sensors have also been reported. The addition of combination of carbon black and multi-walled carbon nanotubes (MWCNT) in epoxy matrix results in a strain sensor with additional capabilities. Recently, Anand and Roy Mahapatra have developed such highly sensitive strain sensor with the addition of 33% carbon black and ~0.57% MWNTs that possess quasi-static and dynamic strain sensing capability, which is common parameter of interest to be monitored in aerospace structures. The addition of carbon black brings down the cost of the sensor. They studied the variation of MWNT concentration on the performance of the strain sensor and brought out a technique to actively tune the gauge factor by varying the bias voltage of the sensing circuit (a voltage divider).

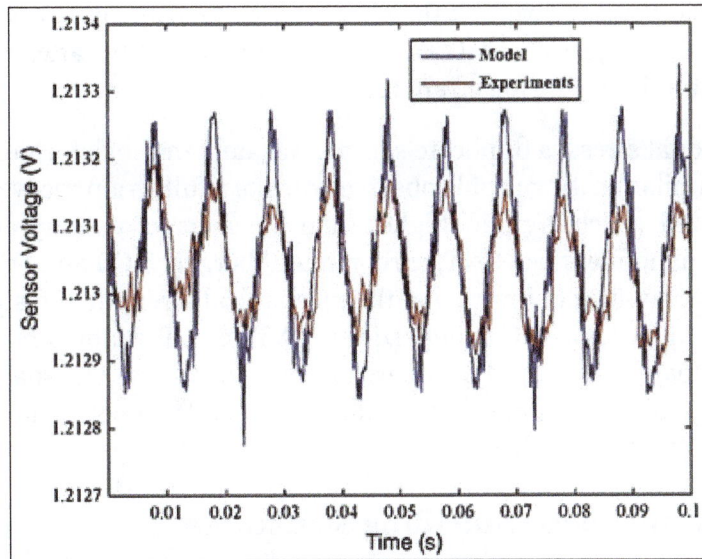

Voltage response predicting the voltage drop caused by reduction in pressure over the PU/CB nanocomposite paint sensor.

Shape Memory Deployable Structures

Shape memory polymers recover strain of over 50–400% upon the application of heat. This ability has attracted the application of shape memory polymers in deployable aerospace structures. Recoverable strain in shape memory polymers is several orders higher than shape memory alloys. However, recovery stress is limited due to their lower stiffness. Recoverable stress improves by the addition of reinforcements demonstrated the storage and release of internal stress in active PMNC with SiC nanoreinforcements. About 20 wt% SiC nanoparticles (700 nm diameter) were added to an epoxy polymer resin. A well-dispersed nanoparticle in polymer matrix was confirmed by SEM as shown in figure. The stress–strain response of the epoxy/SiC nanocomposite at deformation temperature $T_d = 25\,°C$ is shown in figure. Elastic response exhibiting up to 50% strain followed the yielding and inelastic flow. The unloading resulted in an elastic recovery initially, followed by a nonlinear and spontaneous strain recovery. A residual inelastic strain of 35% was fully recovered by heating at $T_R = 120\,°C$.

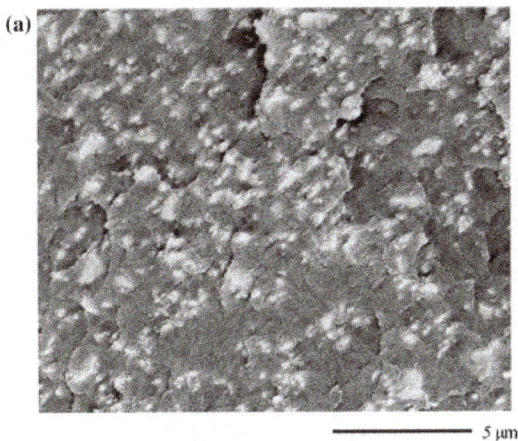

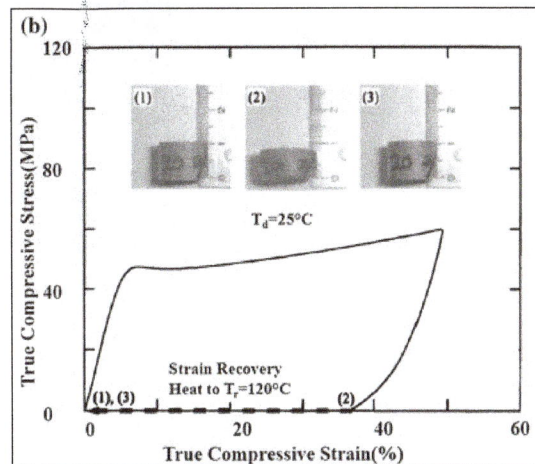

(a) Micrographs showing SEM of epoxy/SiC nanocomposite with 40 wt% SiC. (b) Stress–strain curve under compression and recovery of the nanocomposite shape memory polymer. The insets display the sample shape (1) processed, (2) after compression and (3) after recovery. The sample is free from external conditions during strain recovery.

To determine the residual stress, a duplicate sample was held in the deformed state under a fixed strain subjected to a similar heating profile observed during a full strain recovery. A recovery stress of 31 MPa was observed which is much higher than pure shape memory polymer resin. Other nanocomposites with nanofillers like CNT, carbon nanofiber, clay, CB and other organic particles also possess similar capabilities of improving the mechanical strength and shape recovery stress. Clay imparts similar properties as SiC nanoparticles. CNT and CB nanofillers impart electrical and thermal conductive capabilities required by aerospace applications. The shape recovery property along with the thermal and conductive properties proposes the epoxy/SiC nanocomposites for aerospace deployable structure applications.

Conductive Electromagnetic Shielding Structures

Modern airliners like Boeing 787 and Airbus 380 face challenges to satisfy mechanical and electro-magnetic requirements simultaneously. Electromagnetic (EM) shielding is the ability to reflect and absorb the electromagnetic radiation to act as a shield against the penetration of radiation through it. Apart from providing EM protection from lightning strike, conductive composites offer inten-tional interference with radar-absorbing materials, equipment-level shielding, protection from high-intensity radiated fields and mitigation of human exposure. Carbon fiber composites have been extensively studied for such applications since their properties can be tailored. Epoxy-based nanocomposites have been well proven for their EMI shielding capabilities.

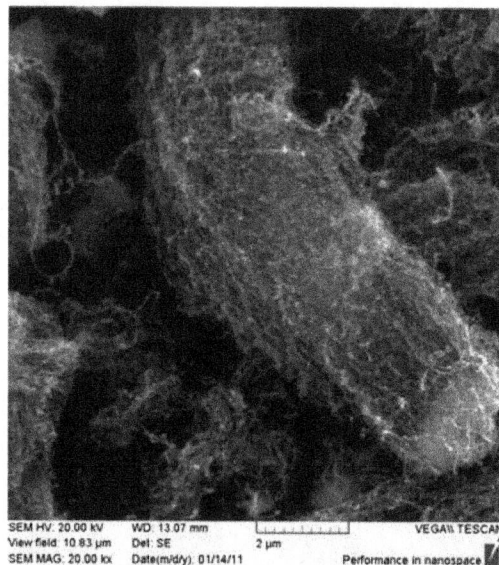

SEM image of epoxy/CNT nanocomposite with 3 wt% MWCNTs.

Compared to nanocomposites with silver, iron oxide (Fe_3O_4) and iron (Fe) nanoparticles, epoxy/CNT nanocomposites have been reported to have better EMI shielding properties. Epoxy/CNT nanocomposites showed 99.9% energy absorption capability in the X-band (8–12.4 GHz). A SEM of epoxy/CNT nanocomposite showed the electrical and thermal paths in the form of CNT net-works. Incorporation of CNT effects the mechanical properties as shown in table. The shielding effectiveness *SE* is defined by the ratio of signal received with the sample and without sample given by,

$$SE(dB) = 10 Log \left[\frac{S_{OPEN}^2}{S_{MATER}^2} \right]$$

where S_{OPEN} is the scattering parameter (that is, the ratio of signal transmitted to signal received by the antenna) in the presence of shielding material and S_{MATER} is the scattering parameter in the absence of shielding material.

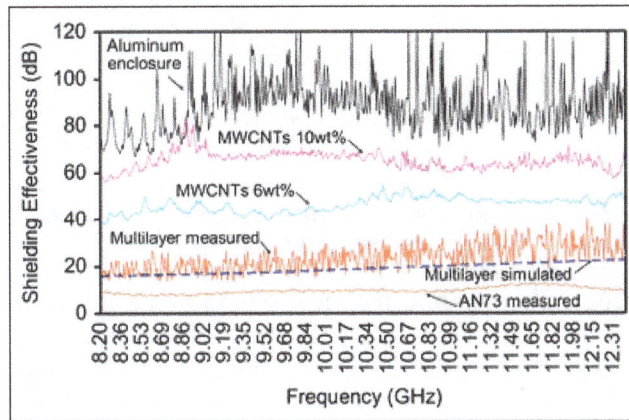

Shielding effectiveness of a shielding structure with single layer of 6 wt% MWCNTs and 10 wt% MWCNTs.

The improvement in the shielding effectiveness measured by a vector network analyzer is shown in figure. The measurements were done with the sample normal to the direction of electromagnetic wave propagation. The measured *SE* matched well with the *SE* predicted by a mathematical model following a winning particle optimization algorithm. The high shielding effectiveness property of 10 wt% CNT nanocomposite was comparable to that of an aluminum enclosure. Addition of multiple layers of shielding materials can give rise to additional reflections, which increases the *SE*. Epoxy/CNT nanocomposites also showed flame retardancy due to the formation of a jammed network structure in polymeric matrix. Moreover, multifunctional properties like high thermal and electrical conductivities and mechanical properties make them suitable for aerospace application where strength, lightning strike protection and EMI shielding are required. EM and radar-absorbing capabilities can be realized using biphase and triphase carbon fiber-reinforced nanocomposites. Thus, these nanocomposites can be installed on aircrafts as advanced EM shields for stealth applications or EMI suppression.

References

- P.M. Ajayan; L.S. Schadler; P.V. Braun (2003). Nanocomposite science and technology. Wiley. ISBN 978-3-527-30359-5

- Lightweight-metal, material-matters: sigmaaldrich.com, Retrieved 11 May, 2019

- B.K.G. Theng "Formation and Properties of Clay Polymer Complexes", Elsevier, NY 1979; ISBN 978-0-444-41706-0

- Flame Retardant Polymer Nanocomposites" A. B. Morgan, C. A. Wilkie (eds.), Wiley, 2007; ISBN 978-0-471-73426-0

Permissions

Index

www.ingramcontent.com/pod-product-compliance
Lightning Source LLC
Chambersburg PA
CBHW082037190326
41458CB00010B/3389